QUANTUM REVOLUTION III
WHAT IS REALITY?

Vignettes in Physics
A Series by G. Venkataraman

Published
The Many Phases of Matter
Why Are Things the Way They Are?
Bose and His Statistics
Chandrasekhar and His Limit
A Hot Story
At the Speed of Light
The Quantum Revolution (3 vols.)
1. The Breakthrough
2. QED: The Jewel of Physics
3. What is Reality?

Forthcoming
Raman and His Effect
Bhabha and His Magnificent Obsessions

Vignettes in Physics

QUANTUM REVOLUTION III
WHAT IS REALITY?

G. Venkataraman

Universities Press

© Universities Press (India) Limited 1994

First published 1994
ISBN 81 7371 004 X

Distributed by
Orient Longman Limited

Registered Office
3-6-272 Himayatnagar, Hyderabad 500 029 (A.P.), India

Other Offices
Kamani Marg, Ballard Estate, Bombay 400 038
17 Chittaranjan Avenue, Calcutta 700 072
160 Anna Salai, Madras 600 002
1/24 Asaf Ali Road, New Delhi 110 002
80/1 Mahatma Gandhi Road, Bangalore 560 001
365 Shahid Nagar, Bhubaneshwar 751 007
3-6-272 Himayatnagar, Hyderabad 500 029
City Centre Ashok, Govind Mitra Road, Patna 800 004
28/31, 15 Ashok Marg, Lucknow 226 001
S. C. Goswami Road, Panbazar, Guwahati 781 001

Typeset by
Access Center, Secunderabad 500 003

Printed in India by
Navya Printers, Hyderabad 500 482

Published by
Universities Press (India) Limited
3-5-820 Hyderguda, Hyderabad 500 029

The cat on the cover was sketched by V. Suresh, whose memory has inspired this series.

Contents

Preface	vii
1. The background	1
2. The EPR paradox	12
3. The cat paradox	22
4. Spin system and the EPR argument	30
5. Test of Bell's inequality	51
6. On measurements	69
7. Complementarity questioned	76
8. Laboratory cousins of Schroedinger's cats	93
9. Where does all this leave us ?	97
Suggestions for further reading	125
Index	127

Preface

To the adult reader

This book and others in this series written by me are inspired by the memory of my son Suresh who left this world soon after completing school. Suresh and I often used to discuss physics. It was then that I introduced him to the celebrated *Feynman Lectures*.

Hans Bethe has described Feynman as the most original scientist of this century. To that perhaps may be added the statement that Feynman was also the most scintillating teacher of physics in this century.

The Feynman Lectures are great but they are at the textbook level and meant for serious reading. Moreover, they are a bit expensive, at least for the average Indian student. It seemed to me that there was scope for small books on diverse topics in physics which would stimulate interest, making at least some of our young students take up later a serious study of physics and reach for the Feynman as well as the Landau classics.

Small books inevitably remind me of Gamow's famous volumes. They were wonderful, and stimulated me to no small extent. Times have changed, physics has grown and we clearly need other books, though written in the same spirit.

In attempting these volumes, I have chosen a style of my own. I have come across many books on popular science where elaborate sentences often tend to obscure the scientific essence. I have therefore opted for simple English, and I don't make any apologies for it. If a simple style was good enough for the great Enrico Fermi, it is also good enough for me. I have also employed at times a chatty style. This is deliberate. Feynman uses this with consummate skill, and I have decided to follow in his footsteps (whether I have succeeded or not, is for readers to say). This book is meant to be read for fun and excitement. It is a book you can even lie down in bed and read. Without going to sleep I hope!

Naturally I have some basic objectives, the most important of which is to stimulate the curiosity of the reader. Here and there the reader may fail to grasp some details, and in fact I have deliberately pitched things a bit high on occasions. But if the reader is able to experience at least in some small measure the *excitement* of science, then my purpose would have been achieved. Apart from excitement, I have also tried to convey that although we might draw boundaries and try to compartmentalise Nature into different subjects, she herself knows no such boundaries. So we can always start anywhere, take a random walk and catch a good glimpse of Nature's glory. Where she is concerned, all topics are 'fashionable'. There is today an unnecessary polarization of the young towards subjects that are supposed to be fashionable. To my mind this is unhealthy, and I have tried to counter it.

This series is essentially meant for the curious. With humility, I would like to regard it as some sort of a 'Junior Feynman Series', if one might call it that. With much love, and sadness, it is dedicated to the memory of Suresh who inspired it.

To the young reader

For over sixty years, we have been merrily using quantum mechanics. In addition to physicists, even chemists, engineers and biologists are being exposed to the subject on account of its great practical value. There are new disciplines called quantum electronics and quantum optics, and in Japan there is even an experimental high-speed train that runs by exploiting superconductivity, a macroscopic quantum phenomenon. All this has tended to create the impression that everything about quantum mechanics is well understood. Not true; in fact, in addition to the doubts raised by Einstein over half a century ago, fresh questions, especially of a philosophical nature, have cropped up. As you will find from this volume, the last word on the subject is yet to be said but even as it is, there is little doubt that quantum mechanics is the greatest revolution witnessed in physical science.

Acknowledgements

Among all the volumes in this series, this has been a particularly difficult one to write. In this effort, I have received (as always) valuable support from Professor N. Mukunda. Particular thanks are due to Professor Partha Ghose and Dr. Dipankar Home, not only for explaining to me in detail their own work but also offering various useful suggestions. Mr. A. Ratnakar provided kind help in gathering source material. In addition, I would like to express my thanks to Professor P.C.W. Davies and to the Cambridge University Press for granting permission to quote from the volume *The Ghost in the Atom* (mentioned in the suggestions for further reading). Mrs. Naga Nirmala played her usual efficient role in producing the manuscript. The friendly and continued cooperation of the publisher is also thankfully acknowledged. I am conscious that there is a larger

Reality that transcends what has been discussed in this book. To that Supreme Reality, my loving salutations.

G. VENKATARAMAN

1 The Background

> Quantum mechanics says that nature is unintelligible except as a calculus, that all you can do is to compute with the equations and operate your apparatus and compare the results.
>
> David Bohm

In Parts I and II, I was busy telling you the story of the discovery of quantum mechanics (QM) and some of its important achievements. But barring a few passing references (in Part I), I said little about the philosophical implications of this new theory. In a sense I dodged the issue because (as you will soon see), the subject is rather difficult. Time now to stop dodging and to squarely face the question!

1.1 What is a scientific theory?

Quantum theory is a scientific theory. Like all other modern scientific theories, it is the result of a long and careful study, in this case of microscopic phenomena.

All scientific theories start from some basic assumptions or postulates. Special relativity, for example, starts with two postulates due to Einstein (see the companion volume *At the Speed of Light*). Starting from such basic postulates, all theories of physics seek to develop a mathematical structure which is also logically consistent. The theory is then ready for use. A good theory not only explains quantitatively all known physical phenomena of the kind it is supposed to deal with, but also makes successful predictions about phenomena not known before—by this I mean that the predictions are verified later by experiments. General relativity, for example, predicted that light would be bent by gravity; this prediction was subsequently verified by experiments.

1.2 Theory as a model for reality

Ultimately, all theories of physics are intended as models for (physical)

2   What is reality?

reality. Sunrise, sunset, motion of the Moon and the planets across the sky—all these are matters of experience for us and can be taken as examples of physical reality. This word reality appears deceptively simple but behind it lie many complications which is why I have to devote a whole book to the subject. For a moment, let us forget all such complications. Ptolemy sought to explain sunrise, sunset, eclipses, etc.,—that is reality relating to astronomy—via a model in which the Earth was at the centre and the planets and the Sun went round the Earth. Today we use instead the Copernican model in which the Sun is at the centre (of the Solar system) rather than the Earth. So the *same* reality was described differently by Ptolemy and by Copernicus. The point I am trying to make is that theories of physics essentially attempt a mathematical description of physical reality. Which theory eventually survives depends of course on which one provides the best description of reality. The question around which this book basically revolves is: *What is the reality that quantum mechanics is trying to describe*?

1.3   Some jargon

At this stage, I would like to introduce you to two fancy words that philosophers often use namely, *epistemology* and *ontology* . According to the dictionary, epistemology means the theory of knowledge; in our context, it simply means OUR knowledge or conception about a physical system. Notice the stress on the word *our*. The emphasis is laid because our knowledge about the system might in fact be quite different from what the system actually is but that is an altogether different matter.

The dictionary meaning of ontology is: branch of philosophy that deals with the nature of being. In our context, it refers to the nature of the system *as it actually exists* . Roughly speaking epistemology deals with knowledge built up from observation whereas ontology refers to *attributes the system has, independent of whether one observes them or not* .

May be I should illustrate the meaning of these two words with an example. Suppose a coin is tossed. You will agree that there is a 50% chance for head and an equal chance for tail. The prediction of the outcome of the tossing is *probabilistic* Now tossing is a mechanical act. So it is legitimate to wonder why one cannot use mechanics all the way and precisely predict the outcome instead of merely giving a probabilistic forecast. True the calculation may be lengthy and

tedious but that is not the point; the point is one of principle.
What are the inputs needed for such a deterministic calculation? We must know the size, shape and the weight of the coin, precise details about the force applied during the toss including which part of the coin was forced and how, etc. Anything else? Yes, we would like to know if there was a breeze and if so, the direction it came from and its velocity. What if there was no breeze? Do we have all the inputs in that case? Not quite because we still do not know how the molecules of air collided with the coin as it was spinning in the air before falling to the ground. Trillions and trillions of molecules would collide with the coin and the bombardment would be completely random. There is little hope of knowing all these complex details, and it is this *ignorance* that forces one to make a probabilistic prediction of the outcome of the toss rather than a deterministic one. Although *we* might be ignorant about the movement of the individual atoms, it could be argued that God would know all about it. So *He* can predict accurately whereas we can assign only probabilities. It's all due to *our* ignorance! In short, the system is intrinsically deterministic but *OUR* knowledge of it is probabilistic. Thus we have here a case of ontic determinism but epistemic indeterminism. I hope it all makes sense!

1.4 Classical physics

By classical physics I refer to physics as it existed prior to Planck's discovery of the quantum of action. Of the many aspects of classical physics, we shall restrict attention here to those which are important when comparisons are made with quantum theory.

The first noteworthy feature is that classical theory always deals with continuously varying quantities; for example, the variables, energy and action always vary continuously in classical physics. By contrast, in quantum theory energy is often quantised and action always is.

Another important feature of classical physics is *determinism* which in turn arises due to causality. In other words, determinism and causality are two sides of the same coin.

1.5 Observables and measurement

Physical systems are characterised by many properties like mass, velocity, position, etc. To determine the values of these properties, one often carries out a suitable measurement.

4 What is reality?

The concept of measurement is very important in quantum mechanics, and so let us try to understand what exactly measurement means and implies, in the first instance in classical physics.

To carry out a measurement, we need an apparatus or an instrument. For example, to measure mass we use a balance. A measurement is usually carried out by bringing the apparatus and the system into some kind of a contact. For instance, to determine the temperature of water in a beaker, one would insert a thermometer (apparatus) into the water (the system).

Two important assumptions made in classical physics in connection with measurements are:

(i) The system already has a definite value for the property being measured, *even before the measurement is actually made.*
(ii) The instrument or the apparatus does not affect the system.

Going back to the example of water in the beaker, what these assumptions mean is that firstly the water had a definite temperature even before the thermometer was inserted, and secondly the thermometer does not affect the value of the temperature.

It might happen that as a result of using a bad instrument, the value sought to be measured is affected. For example, a bad current meter would give an incorrect reading by consuming some current itself. However, by using a good meter, this perturbation can be minimised. It is therefore assumed that by proper design, one can in principle atleast, reduce to zero the interaction between the apparatus and the system. The system that is being observed is thus not disturbed in any way by the measurement and the correct value for the property is obtained. This is also the value the property had *before* the measurement was carried out. Thus in classical physics the physical properties are attributes of the system which exist on their own, whether or not someone takes the trouble of measuring them. This last point is very important, and we shall come back to it later.

1.6 Indeterminism in classical physics

Classical physics is often described as deterministic. While this might be true when one is dealing with just a few particles, in the case of very large numbers (as in statistical mechanics), probabilistic considerations have to be brought in (see the companion volume *A Hot Story*). The important point to note here is that indeterminism enters on account of *OUR* ignorance or incomplete knowledge about the positions and the velocities of the molecules. But the Superior Being knows all these details, and so there is no ontic indeterminism. By contrast, in quantum mechanics there is an underlying indeterminism even in the case of a single particle. More about all this shortly.

1.7   The quantum world

We enter now the world of quantum physics. As I narrated in Part I, quantum mechanics was discovered while trying to describe microscopic processes, e.g., the behaviour of electrons in atoms. At first people tried to use classical mechanics and classical electrodynamics for this purpose but this did not work. Then followed a period when people limped along with models. This model-based approach worked sometimes and failed on other occasions. And then in two short years during the mid-twenties, a powerful tool kit to deal with (non-relativistic) problems of microscopic physics was assembled, thanks to pioneering contributions from Heisenberg, Schroedinger, Dirac, Born and a few others. Soon special relativity was brought into the picture, and roughly a couple of decades later, quantum electrodynamics also was given a good shape (at least at the working level). A glimpse of all this has been given in the first two parts.

I wish to discuss now the *physical* content of quantum mechanics, what it means and implies, and in what essential way it differs from classical physics. We start with Landau's remark that among the theories of physics, the quantum theory occupies a unique place by virtue of a peculiar double role it plays. In physics, a theory is often superseded by a more general and a more complete one. Usually the more general theory can be formulated in a logically complete manner all by itself, *without* any reference to the simpler theory it overshadows. For example, special relativity can be formulated without any reference to Newtonian mechanics. However, once a general theory is developed, one could obtain the results of the simpler theory by a suitable limiting process. Thus in the case of special relativity, the results of Newtonian mechanics can be obtained by considering the limit $v \ll c$.

In the case of quantum mechanics, things are quite different. It is true that the results of classical mechanics can be recovered from quantum mechanics by a suitable limiting process. But there is one respect in which QM is different from all other "general" theories of physics which is that it is impossible *even in principle*, to formulate the basic concepts of quantum mechanics *without* referring to classical physics. This last remark might appear mystifying at present but should become clear later.

1.8 QM—a gist

The essence of quantum mechanics can be summarised by the following statements (recall Part I):

- A quantum system is supposed to have "observables"
- In the mathematical theory, observables are represented by operators.
- Every measurement of an observable yields one of the eigenvalues of the corresponding operator.
- If $\hat{A}$ and $\hat{B}$ are two operators which commute, i.e., $\hat{A}\hat{B} - \hat{B}\hat{A} = 0$, then the corresponding observables can be measured simultaneously with as much accuracy as we want (meaning that one could in principle not only measure $\hat{A}$ and $\hat{B}$ simultaneously, but also with infinite precision). Since the position operator $\hat{q}$ and the momentum operator $\hat{p}$ do not commute with each other, they cannot be measured simultaneously. This is an echo of the uncertainty principle.
- A quantum mechanical state is represented by a vector in a suitable space.
- If the system is in a definite state $|\Psi\rangle$ and if observations of an observable $\hat{O}$ are repeatedly made with the system in this state, then the average result would be equal to $\langle\Psi|\hat{O}|\Psi\rangle$ (assuming $|\Psi\rangle$ is normalised).
- A measurement always causes the system to jump into an eigenstate of the dynamical variable that is being observed.

A number of delicate points have been mentioned above and at least some of them require further elaboration.

1.9 Measurement in QM

Measurements play a crucial role in QM because the value one obtains for a particular observable is the result of the measurement itself. Thus if one tries to measure the coordinate of an electron, the value one would obtain is due to the perturbations caused by the **measurement** process itself. I have explained this in Part I while discussing the gamma-ray microscope. It is important to appreciate that it is not possible *even in principle* to reduce to zero the disturbing influence produced by the measurement process. This is in marked contrast to classical physics. In fact if the disturbance due to the measuring apparatus can be made zero (as in classical physics), then it would mean that the measured quantity has itself a definite value independent of the measurement. QM rejects this view.

Figure 1.1 amplifies the meaning. We have an atom which makes a transition between two of its levels. We would immediately say that

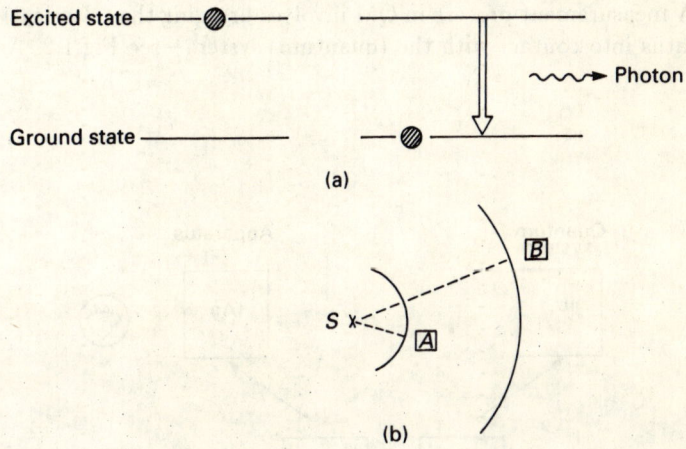

Fig.1.1(a) illustrates the emission of a photon due to a transition between the levels of an atom. In space, the electromagnetic field would propagate as a wave field as shown in (b). Photon detectors like A and B would detect photons localised at A or B as the case may be. When a detection occurs, the wave collapses. We can then identify the photon trajectory SA or SB.

a photon is emitted. But we must be careful in making such a statement. We should not imagine that the atom is shooting a photon rather like a bullet fired from a gun. If we don't have an apparatus to detect the emitted radiation but would merely like to give a theoretical description of the emission, then we would describe it via quantum field theory. Loosely speaking, it is as if a spherical wave is spreading from the atom. (Caution: This refers to a probability amplitude wave, and should not be confused with the wave of classical optics.) Suppose we place a photon detector at A and that this detector detects a photon. At the instant of detection, the spherical probability amplitude wave *collapses* to a local wave packet. In a

8 What is reality?

manner of speaking, the act of observation creates a photon at the space point A. If instead we had placed the detector at B and this detector registers, the photon would have been created at B.

In short, without an observation, the value of a dynamical variable cannot be given any meaning in QM. Thus QM without measurement is meaningless. What about the apparatus? Is that also a quantum object? It had better not be for if it were, we would be in all kinds of trouble! In fact the apparatus is taken to be a classical object, capable in principle of infinite precision. This is how classical physics becomes a foundation and an indispensable part of QM.

A measurement process in QM involves bringing the (classical) apparatus into contact with the (quantum) system—see Fig.1.2. As a

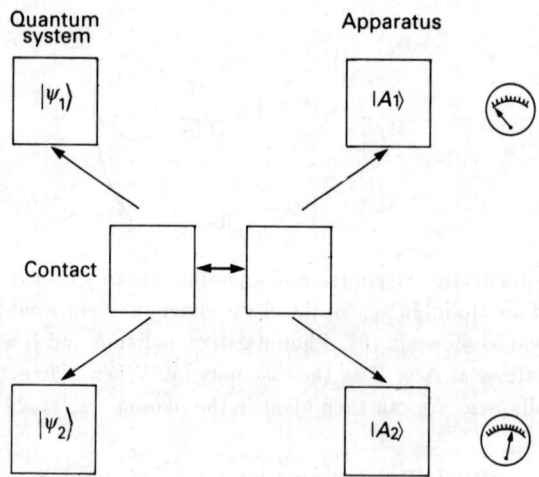

Fig.1.2 Schematic of the measurement process in QM. Initially, the quantum system is in a state $|\Psi_1\rangle$ which is a mixture of potentialities. The contact of the apparatus with the system causes one of the potentialities to become reality, in turn producing the change $|\Psi_1\rangle \rightarrow |\Psi_2\rangle$. For convenience, we represent the states of the apparatus also using the Dirac notation, though it is a classical object. From the change ($|A_1\rangle \rightarrow |A_2\rangle$), symbolised by the meter readings, one obtains information about the state $|\Psi_2\rangle$.

result, both the apparatus and the system being observed change their states. From the reading corresponding to the apparatus state $\mid A_2\rangle$, we infer something about the state $\mid \Psi_2\rangle$ of the system. Note: We learn nothing about the quantum state $\mid \Psi_1\rangle$ that existed *before* the apparatus and the system came into contact. We can at best learn only something about the state produced *after* the interaction between the apparatus and the system. More will be said later about measurement, apparatus, etc.

1.10 Complementarity

A good bit of what is to be discussed in this book arises out of remarks made in the previous section. Some of the concepts mentioned there are quite contrary to normal experience and therefore when quantum mechanics first made its appearance, these concepts not only caused a lot of confusion but also much uneasiness. Many people simply did not know what to make of it all, although they knew how to use the formulae of quantum mechanics and make calculations.

One person who pondered a lot about what quantum mechanics actually means and implies was Niels Bohr. He will make his appearance again in the next chapter but here let me give a glimpse of the interpretation he gave to quantum mechanics. This interpretation is often referred to as the *Copenhagen intepretation*.

In simple terms, what Bohr said was the following: A quantum system intrinsically has only *potentialities* . For example, an electron has the potentiality to appear or behave like a particle or to behave like a wave. It has the potentiality to have a particular value p for the momentum or the potentiality to have a value q for the coordinate. Similarly, electromagnetic radiation has the potentiality to act like a wave or a light quantum, i.e., a photon. About potentialities Heisenberg once declared, "Atoms or elementary particles are not real; they form a world of potentialities or possibilities rather than one of things or facts."

In all the examples cited there are two potentialities and they are associated with *complementary* aspects e.g., a wave-like aspect or a particle-like aspect. When a measurement is made, only one of these aspects is actually realised. The other complementary aspect becomes irrelevant, and it does not make any sense at all to talk about it. So the outcome of a measurement depends on what we choose to observe. Pascual Jordan who has made important contributions to QM once remarked that observations not only *disturb* what has to be measured but in fact *produce* it. As he put it,

The electron is forced to a decision. We compel it *to assume a definite position*; previously it was, in general neither here nor there; it had not yet made its decision for a definite position ... If by another experiment the *velocity* of the electron is being measured, this means: the electron is compelled to decide for itself some exactly defined value for the velocity ... we ourselves produce the results of the measurement.

It is important to get one thing very clear. According to quantum mechanics, prior to a measurement *we cannot even think* that the electron has a definite momentum and a definite coordinate. As Bohr says (effectively): "Before an actual measurement, an electron has no definite momentum nor any definite coordinate. It has only a possibility of having or a potential to have a momentum p say, and a potential to have a position coordinate q say. Once we try to observe the electron (with a classical apparatus of course), one of the many potentialities gets converted into a reality. Till that instant, it does not make any sense to talk of a reality associated with the position or the momentum of the electron. Before one talks of reality, one must specify the experimental arrangement one is going to employ." (Don't confuse the word potential with potentials like the electrostatic potential or the gravitational potential. I introduced it because some experts like to talk of quantum potentialities.)

But this is not all. Just as one swallow does not a summer make, the observation of one individual event need not necessarily have an objective meaning. Rather, the experiment must be repeated again and again with the same initial conditions to get a meaningful answer. The only meaningful answer is a quantum average. This too is of profound significance because one can obtain only *statistical* information about the microscopic world. In turn this means that one must give up classical notions of determinism and causality (recall Box 8.1 in Part I).

If you seriously think about all this, it is a shattering pronouncement, completely contrary to normal everyday experience and quite against what everybody including philosophers believed in from ancient times. No wonder so many people balked at the whole idea. But then, as I pointed out in Part I, one by one the leading lights like Ehrenfest, Born, etc., fell in line, the sole exception being Einstein. Even he yielded after a while, though only partially. Everyone was forced to accept quantum mechanics, however unpleasant it was from a philosophical point of view because quantum mechanics alone could satisfactorily explain microscopic processes.

1.11 Classical versus quantum physics

Let us wrap up this chapter with a few remarks about the crucial differences between classical and quantum physics. Firstly, of course, classical physics can stand all by itself whereas quantum physics needs classical physics as a foundation. (It is a different matter that classical physics does not always work.) Secondly, in classical physics measurement need not disturb the system whereas in quantum physics it does. Thirdly, classical physics is built upon the concepts of causality and determinism whereas in quantum physics one has only statistical determinism. Lastly, in classical physics one has what is called *objective reality*, i.e., the world has a definite state of existence independent of whether we observe it or not whereas in QM the actual state of existence depends in part on how we observe and what we choose to observe.

As I said earlier, in the early days of quantum mechanics all these differences left many people quite uneasy but eventually everyone fell in line with Bohr's thinking. One person who refused to toe the Copenhagen line was Einstein. To start with he argued that quantum mechanics itself was wrong because it denied determinism (which he felt was a necessity). I have already narrated in Part I how Einstein tried to construct various *gedanken* (thought) experiments to refute quantum mechanics. But Bohr successfully demolished all those objections. By around 1930 or so, Einstein reconciled himself to the view that quantum mechanics was a good working prescription and that it was able to correctly predict the outcome of experiments in the microscopic world. He stopped arguing that QM was wrong. Instead he decided to attack QM's rejection of objective reality. We pick up that story in the next chapter.

2   *The EPR Paradox*

> No content can be grasped without a formal frame and that any form, however useful it has hitherto proved, may be found to be too narrow to comprehend new experience.
>
> *Niels Bohr*

The early thirties were years of great turmoil in Germany. Adolf Hitler had seized power and launched a vicious persecution of people of Jewish origin. Among other things, this led to an exodus of eminent Jewish scientists. After an exploratory trip to America in 1930, Einstein left with bag and baggage in 1933 to take up an appointment at the Institute for Advanced Studies, Princeton. Personal safety was no longer a problem, and physics once again began to receive his attention.

2.1   The EPR collaboration

Starting from around 1920, Einstein was engaged in an attempt to unify general relativity and electromagnetism into one unified theory. Once he settled down in Princeton, Einstein went back to this programme. In 1934, a visiting scientist at the Institute named Nathan Rosen posed a question about quantum mechanics to Einstein. This got him (Einstein) thinking about his old doubts and reservations. The result was a joint paper with (Boris) Podolsky and Rosen as collaborators, which has since become famous as the EPR paper.

In the confrontation between Bohr and Einstein that occurred in 1930, Bohr emerged the victor with Einstein conceding that QM was not wrong and that it worked. He even said that "the Born statistical interpretation of the quantum theory is the only possible one," and that "statistical quantum theory is the most successful theory of our period." In other words, distasteful though it was to him, Einstein reconciled himself to giving up determinism. But where reality was concerned, he would not give up so easily. He strongly felt that QM did a poor job of describing reality and that a future theory would have to set matters right. In a lecture delivered during that period he declared:

> I still believe in the possibility of giving a model of reality which shall represent events themselves and not the probability of their occurrence.

The EPR collaboration provided an opportunity to highlight the limitations of QM in describing physical reality, notwithstanding its enormous success as a prediction machine. Thus QM was at best a successful theory but not a complete one.

2.2 EPR's objective

The EPR paper was titled: *Can Quantum Mechanical Description of Physical Reality be Considered Complete?* EPR start with the conviction that objective reality exists, whether we observe it or not. It does not depend on our measurements or even on our own existence.

Getting down to brasstacks, one must have a definition of reality so that one could analyse whether or not QM provides a proper description of physical reality. EPR propose the following criterion:

> If, without in any way disturbing a system, we can predict with certainty (i.e., with probability equal to unity) the value of a physical quantity, then there *exists an element of physical reality corresponding to this physical quantity.*

EPR remark that in judging the success of a physical theory, two questions must be answered: "(i) Is the theory correct? (ii) Is the description given by the theory complete?" As far as the first question is concerned, EPR do not doubt the correctness of QM. As I mentioned earlier, Einstein had conceded this point soon after his debate with Bohr. The point at issue now was whether QM could claim to be a complete theory in the sense of providing a complete description of reality, the criterion for reality being as given above.

All along, the supporters of QM had claimed that QM was a complete theory but nobody had subjected it to a careful test. EPR declared that they proposed "to show, however, that this assumption [that QM is a complete theory] together with [our] criterion for reality, leads to a contradiction." EPR then describe a gedanken experiment which, according to them, brings out this contradiction.

2.3 EPR's idea

I shall now give a simplified description of EPR's gedanken experiment. Suppose two identical particles are held together in tension by a spring as in Fig. 2.1. If the spring is cut, the particles would fly off in opposite directions. The cluster of two particles could be a hydrogen molecule which is ripped apart by an impinging atom (which plays the role of scissors). Consider the situation when the particles are far apart, say 1 km or even 100 km from each other. Denote their

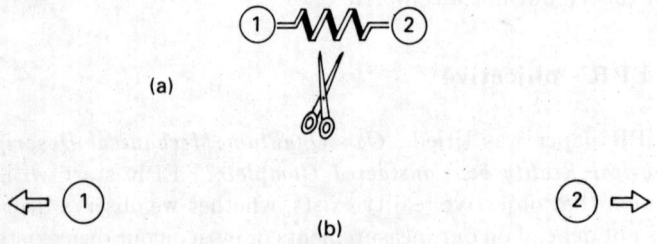

Fig.2.1(a) shows two particles bound together by a spring held in compression. When the spring is cut, the particles would fly off in opposite directions as in (b). EPR describe how the position and the momentum of particle 2 can be deduced purely from observations on particle 1, in violation of the rules of QM.

positions and momenta by (q_1, p_1) and (q_2, p_2) respectively. The total momentum is defined by $P = p_1 + p_2$ while $Q = q_1 - q_2$ denotes the separation or the relative distance. Quantum mechanics imposes the following commutation rules:

$$[p_1, q_1] = -i\hbar \qquad (2.1)$$

$$[p_2, q_2] = -i\hbar \qquad (2.2)$$

$$[p_1, q_2] = [p_2, q_1] = 0. \qquad (2.3)$$

Using these relations, it is easy to show that P and Q commute which means that according to QM they can be simultaneously measured. Note that P is conserved and since the particles are at rest to start with, $P = 0$.

Let's now get to work. We first measure p_1 and P. Then we can determine p_2 since $p_2 = P - p_1$. Thus we have been able to determine

p_2 by observing particle 1 and *without disturbing particle* 2. Recall now the criterion adopted for reality. If the value of a physical quantity can be predicted with certainty without disturbing the system, then there is a reality associated with that quantity. In this case, we have determined p_2 without disturbing particle 2. So p_2 is an element of reality (according to EPR, that is).

Having taken care of p_2, we next measure q_1 and Q. This then enables us to predict the value of q_2, without disturbing particle 2. Therefore q_2 also becomes an element of physical reality. The net outcome is that we find p_2 and q_2 are simultaneous elements of reality. On the other hand, according to QM p_2 and q_2 *cannot* be simultaneous elements of reality because they cannot be simultaneously measured — see (2.2). But through our gedanken experiment we have been able to establish a reality that QM denies. So QM cannot claim to be a complete theory — this in essence is EPR's argument.

EPR are quite careful and point out that holes can be picked in their arguments by demanding that the criterion for reality should be changed. One might argue as EPR themselves suggest "that two physical quantities can be regarded as simultaneous elements of reality only when they can be simultaneously measured or predicted." EPR however feel that this would be too restrictive a definition of reality, besides being an unreasonable one. After all in their experiments they are merely disturbing particle 1 which is 1 km or even more from particle 2. What does it matter if p_1 and q_1 are measured one after the other instead of simultaneously? It is too much to demand that the "reality of p_2 and q_2 depend upon the process of measurement carried out on the first system which does not disturb the second system in any way."

EPR conclude by remarking that while they have shown that

> the wave function does not provide a complete description of the physical reality, we [have] left open the question of whether or not such a description exists.

OK, so Einstein did not believe that the wave function ψ provided a description of physical reality. What precisely were his views on QM? As I have already mentioned, by 1935, Einstein accepted QM but only a makeshift theory that "worked". He said:

> The ψ function does not in any way describe a condition which could be that of a single system; it relates rather to many systems, to "an ensemble of systems" in the sense of statistical mechanics.

This is the so-called *ensemble interpretation*, at least one version of it. Basically it implies that the wave function describes *an ensemble of identically prepared systems* (see last para of section 1.10).

Till the end of his life, Einstein was strongly convinced that QM was not the last word and that it should be possible to construct a classical theory which besides being deterministic, would provide a complete description of physical reality, i.e., objective reality. In a later chapter I shall come back to such theories and the fate they have suffered.

2.4 Enter Bohr

The EPR paper was submitted to the *Physical Review* (a well-known journal published in America) on 25 March 1935 and was published a few months later. As soon as it appeared in print, the famous newspaper *The New York Times* carried a story under the heading "Einstein attacks Quantum Theory". Einstein strongly deprecated the newspaper article. Be that as it may, Einstein *had* challenged QM. Obviously, Bohr the high priest of QM could not let it pass. Rosenfeld, a close associate of Bohr recalls:

> This onslaught came down upon us as a bolt from the blue ... A new worry could not have come at a less propitious time. Yet, as soon as Bohr heard my report of Einstein's arguments, everything else was abandoned.

Later Bohr himself said:

> Due to the lucidity and apparently incontestable character of the argument, the paper of Einstein, Podolsky and Rosen created a stir among physicists and has played a large role in general philosophical discussion. Certainly the issue is of a very subtle character and suited to emphasise how far, in quantum theory, we are beyond the reach of pictorial visualisation [recall the pseudo-realistic pictures prepared by Bohr during his debates with Einstein — Chapter 8, Part I].

Bohr replied to EPR via a paper in the *Physical Review* which was submitted soon after the EPR paper appeared. Interestingly, Bohr used for his article the same title as was used by EPR. Bohr's paper is quite difficult to follow — at least I find it so, and mercifully I am not the only one. The distinguished physicist Bell (about whose contributions you will hear later) also had difficulty in understanding Bohr.

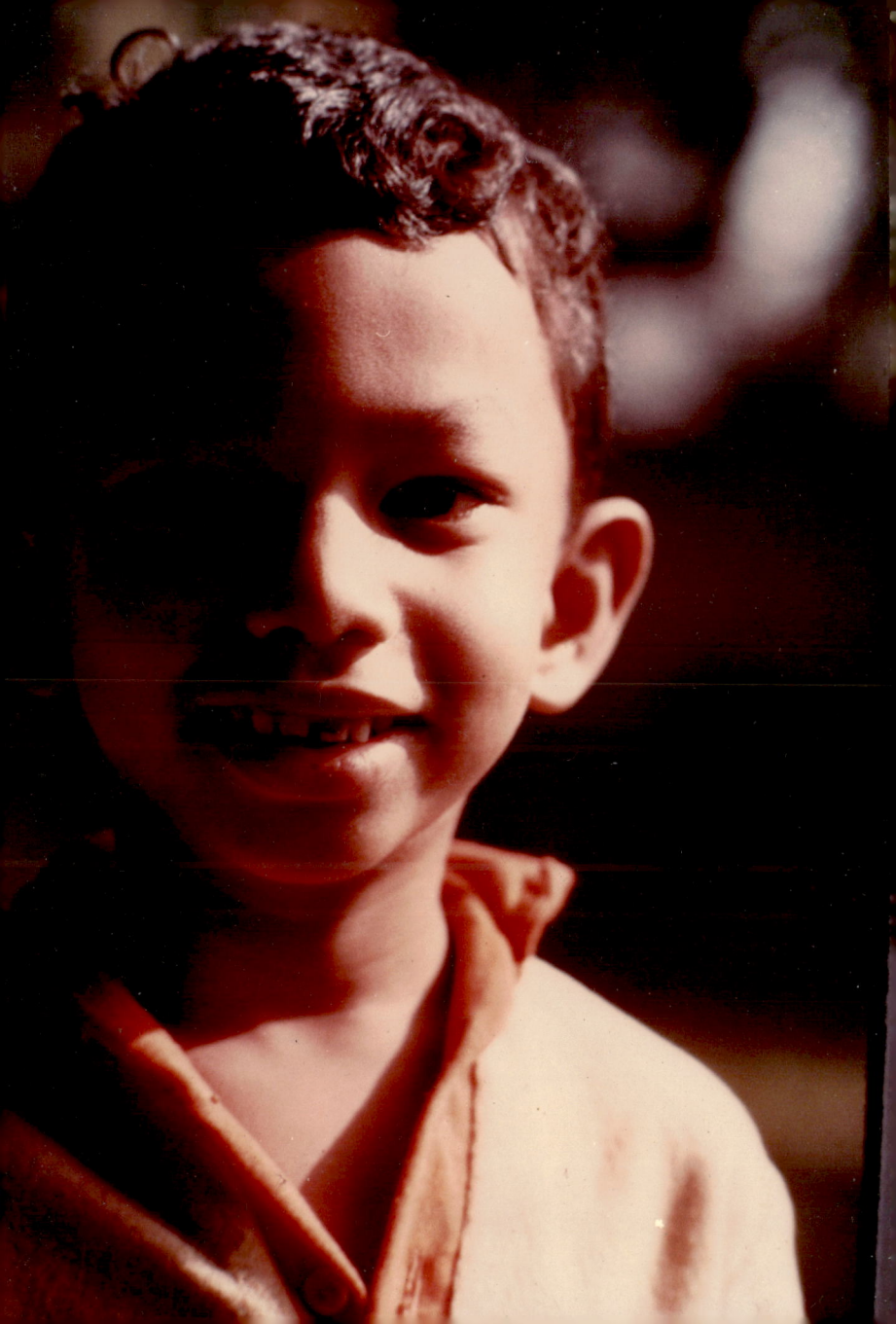

Bohr's style is pretty complicated and when such a style is used for discussing a difficult problem, you can imagine what the plight of ordinary mortals (like myself) would be! In case you think I am unfairly accusing Bohr, here is a sample from the above-mentioned paper.

> This last circumstance is in fact quite analogous to the renunciation of the control of the momentum of the fixed diaphragm in the experimental arrangement discussed above, and depends in the last resort on the claim of a purely classical account of the apparatus, which implies the necessity of allowing a latitude corresponding to the quantum-mechanical uncertainty relations in our description of their behaviour.

Now do you agree with my complaint?! Incidentally, here are two interesting statements from Bohr: "The opposite of a deep truth is also a deep truth!", "Truth and clarity are complementary." I can't resist slipping in Einstein's remark that Bohr thought very clearly, wrote obscurely and regarded himself as a prophet. I wonder what Bohr's response to that was!

Despite my grumbling, Bohr's paper occupies an important place in the history of quantum mechanics because it once again raised the flag for quantum mechanics. Bohr argued that QM is as complete as one could have it and that no change was required in the existing structure of the theory.

The term paradox is defined as: A plausible argument leading from plausible premises to an implausible conclusion. In the literature, there is frequent reference to the EPR paradox even though in their paper EPR themselves do not talk about any paradox. They simply say (in effect): "QM is not able to describe reality and is therefore an incomplete theory." The term paradox was first used by Einstein in a letter to Schroedinger who picked it up and referred to the EPR paradox; the name then stuck.

Bohr's main criticism of the EPR argument was that their criterion for reality was "ambiguous". He wondered what exactly EPR meant by saying "without in any way disturbing the system?" Forget about two particles for a moment, and just consider one. Say we wish to know the momentum of the particle. According to QM, we must use a suitable apparatus to determine that quantity. If we want to know the position then again a measurement is necessary, using a suitable apparatus of course. The experimental arrangements needed in the two cases would be quite different because momentum and position are conjugate variables in QM. Consider now the measurement of position. We could localise the particle by using a narrow slit. However, the slit which is a part of the apparatus must be *rigidly* fixed to the frame of reference containing the particle. Otherwise, when the particle passes through the slit, the latter would be disturbed and the accuracy of our position measurement would suffer. Now when the particle passes through the slit it transfers some momentum to it. But since the slit is rigidly fixed to the frame of reference, the momentum is transferred to the latter. Let us bear this in mind.

Consider now the two-particle system. Even though the two particles might be 1 km apart, we still regard the two as making up a system; otherwise we have no business talking about the total momentum etc. But notice that just because we have a system, it does not mean that particle 1 has to exert a physical force on particle 2. OK, so we have this system of two particles, defined with respect to a particular frame of reference. As EPR want, we set up an experimental arrangement to measure p_1. Fine. EPR now want us to measure q_1. So we dismantle the previous apparatus and set up a new one to measure position. This calls for a slit *rigidly* fixed to the frame of reference with respect to which the system is defined. But now there is a delicate point. When q_1 is being measured, some momentum would be transferred to the slit. Because the slit is *rigidly* fixed to the frame of reference, this momentum disturbance is passed on to the reference frame as a whole. So the momentum of the system too is disturbed and it is no longer P. Therefore, by the time we are through with measuring q_1, we have changed the total momentum from its original value P which means that the value we deduced earlier for p_2 is no longer valid. In other words, p_2 ceases to have reality when q_2 is determined. At a given time therefore, only one of the two quantities has a reality and indeed this reality is what emerges after the measurement (of p_1 or q_1 as the case may be). EPR therefore cannot bypass the constraints of QM by trying to observe particle 1 and claiming that particle 1 cannot influence particle 2. Bohr's punch line is that we must stop thinking about objective reality and radically revise our views about what reality actually stands for.

2.5 Food for thought

Bohr no doubt saved the day for QM but, as you will find from this book, it did not put an end to discussion. Many subtle points came to the forefront and they have been engaging the attention of experts ever since. The first thing we must appreciate is that Bohr did not reject the notion of reality; rather, he objected to the classical version of it. And according to the latter, a particle can have attributes or properties even though they have not been measured. But Bohr maintained that if no observations are made, a particle cannot have *physically real* attributes. To put it all rather simply, the question is: How is (physical) reality realised? And the answer (according to Bohr) is: Reality happens when we look, and what happens depends on *how* we look!

One topic of interest concerns measurements. This is going to occupy us later but here let me take a minute off to say a few words on this subject. You might have noticed that the slit used (by Bohr) for measuring q_1 was made a part of the quantum system and not the apparatus. Normally, we would think of the slit as a part of the (classical) apparatus but here we have treated it as a part of the system being observed. The question therefore is: How does one draw a line and say, "This is the apparatus and that is the quantum system"? This is a tricky point which will keep us busy later.

We now come to the subtle question of *quantum correlations* or *quantum entanglements* as Schroedinger referred to it. In our case, particles 1 and 2 are not mechanically coupled but are always correlated, even though they might be separated by macroscopic distances. According to quantum mechanics (i.e., Bohr!), till a measurement is made we can only talk of a total wave function for the system of two particles, no matter how far apart they are. This is entanglement. When we do a measurement, the measuring apparatus interacts with the *whole* system, i.e., the system of two particles, even though we might think we are dealing with just one particle alone. Consequent to the measurement there is disentanglement, and the wave function of the system now becomes a mere product of the separate wave functions of the two particles.

One sobering thought that came out of all this argumentation was that words that we commonly use have to be used with care when dealing with delicate matters like those under discussion. In fact Bohr himself said:

> Phrases often found in the physical literature, as "disturbance of phenomena by observation" or "creation of physical attributes of objects by measurements" represent a use of words like "phenomena" and "observation" as well as "attribute" and "measurement" which is hardly compatible with common usage and practical definition and, therefore, apt to cause confusion.

Bohr went so far as to urge that the word phenomenon be used exclusively to refer to observations made under specified conditions, with a full description of the circumstances of the experiment and indeed the whole experiment!

I would also like to draw your attention to the fact that the EPR experiment raises QM from the microscopic world to the macroscopic because we are now talking of quantum effects over distances of 1 km or even larger. Bohr once wrote: "Anyone who is not shocked by quantum theory has not understood it." EPR argued that QM is even more shocking than one originally thought it was!

2.6 Einstein's response

How did Einstein respond to Bohr's rebuttal? This time Einstein did not yield. On the face of it this might appear irrational. Not quite. He had a point of view and it cannot be lightly dismissed. Abraham Pais, Einstein's scientific biographer remarks that Einstein wanted things out there to have properties whether or not they were measured. That was his stand and his philosophy. Pais adds:

> We often discussed his notions on objective reality. I recall that during one walk Einstein suddenly stopped, turned to me and asked whether I really believed that the moon exists only when I look at it.

Well, there you have it. What did Einstein think of quantum correlations? He did not like the idea at all and referred to it as "spooky action at a distance". As he put it:

> That which exists in [region] B should ... not depend on what kind of measurement is carried out in part of space A; it should also be independent of whether or not any measurement at all is carried out in space A. If one adheres to this program, one can hardly consider the quantum-theoretical description as a complete representation of the physically real. If one tries to do so in spite of this, one has to assume that the physically real in B suffers a sudden change as a result of a measurement in A. My instinct for physics bristles at this.

Einstein point was simply this: Say A and B are far apart. A measurement is made in A. It immediately produces an effect at B, though B is far away. This implies some signal or information has travelled instantaneously from A to B, which of course is counter to relativity.

The EPR paradox   21

Einstein realised that he was rather lonely in holding on to such views but did not care. A year before he passed away he said: "I must seem like an ostrich who forever buries his head in the relativistic sand in order not to face the evil quanta." He remained so till the end.

Let me end this chapter with a remark made by Bohr on this subject. He was once asked whether there was such a thing as quantum reality which QM tried to portray. In other words, did the quantum mechanical algorithm somehow mirror an underlying quantum reality? (Notice this new word – quantum reality). Bohr replied:

> There is no quantum world. There is only an abstract quantum mechanical description. It is wrong to think that the task of physics is to find out how Nature *is*. Physics concerns what we can say about Nature.

Are you wondering what this reality is that I am trying to write about? I wouldn't blame you!

3 The Cat Paradox

"You'll see me there," said the Cat, and vanished.

Lewis Carroll
in Alice in Wonderland

Einstein was the foremost among those who refused to accept the philosophy of QM but he was not the only one. Schroedinger was another of those who objected, and like EPR he too came up with a paradox of his own; later this became famous as the *cat paradox*. Today the cat paradox is hardly remembered (unlike the EPR paradox) but in its day it produced a sensation of sorts. When you read about it, you will understand why people were disturbed by the paradox.

As I narrated in Part I, Schroedinger came to QM via Hamiltonian optics and Hamiltonian mechanics. From the beginning he was opposed to the probabilistic interpretation, complementarity, quantum jumps and all that. In fact, you might recall from Part I the story of his being cornered by Bohr on the issue of jumps, how he (Schroedinger) fell sick and finally exclaimed, "If we are going to stick to those damned quantum jumps, then I regret that I ever had anything to do with the quantum theory."

Schroedinger entered the fray with a series of papers raising a philosophical objection to the standard interpretation of quantum mechanics, i.e., the Copenhagen interpretation (CI) mentioned in Chapter 1. To highlight the problems underlying the EPR paradox, Schroedinger constructed the following example. There is a scholar who is asked a question about his position or momentum coordinate. This fellow is always prepared to give the correct answer to the first question he is asked but thereafter he becomes tired or disconcerted and all his answers are wrong. However, since he always provides the right answer to the first question he is asked—and he does not know whether he is going to be asked about the position or about the momentum—he must know both the answers! I am sure you see echoes of the EPR argument.

The remark has often been made that the wave function is a model for reality. In 1935, Schroedinger decided to examine this. For this purpose, he now devised a macabre gedanken experiment which is carried out as follows: One first gets hold of a box. Into this is fitted a deadly contraption consisting of a bottle containing cyanide and a hammer held by a clip—see Fig.3.1 and also Box 3.1. The clip is connected to a device that can detect radioactive emission. In front of the detector is placed a radioactive source. The idea is that when

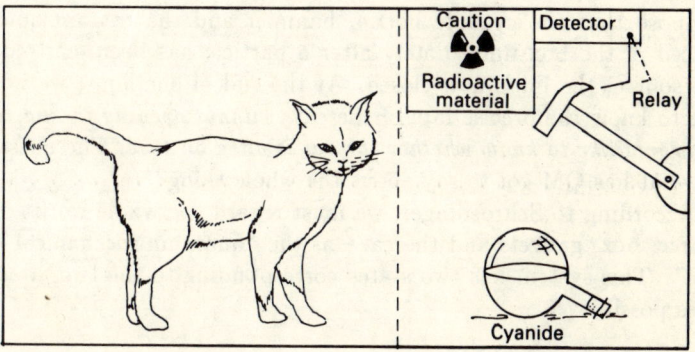

Fig.3.1 Schematic of Schroedinger's gedanken experiment. When the detector detects a particle emitted by the source, the clip is released, the hammer falls, the bottle breaks and the cyanide is released. The cat is then instantly killed. For the full story, see the text.

Box 3.1 This is how Schroedinger himself described the cat experiment.

"A cat is penned up in a steel chamber, along with the following diabolical device (which must be secured against direct interference by the cat): in a Geiger counter there is a tiny bit of radioactive substance, so small, that perhaps in the course of one hour one of the atoms decays, but also with equal probability, perhaps none; if it happens, the counter tube discharges and through a relay releases a hammer which shatters a small flask of hydrocyanic acid. If one has left this entire system to itself for an hour, one would say that the cat lives if meanwhile no atom has decayed. The first atomic decay would have poisoned it."

---

the detector registers a particle emitted by the source, the clip is released, the hammer falls and the cyanide bottle is broken.

Why all this deadly plotting? Ah, that's where the poor cat comes into the picture. When everything is set up and ready, we put the cat into the box and close the lid, in a sense sealing its fate. The instant the source emits a particle, the bottle is broken and the cyanide vapour at once kills the cat. What has all this horrible business got to do with QM? Wait!

Let us suppose that the source is weak and that on the average there is only a 50% chance for a particle to be emitted in an hour. This means that if the source has just emitted a particle then one hour later there is a 50% chance another particle would be emitted. Suppose that the cyanide bottle, hammer and the cat are quickly placed in the box immediately after a particle has been emitted by the source; the lid is now closed. At the end of one hour, we would like to know the precise fate of the cat, *without opening the lid, i.e., we would like to know whether the cat is alive or dead*. The question is, what has QM got to say about the whole thing?

According to Schroedinger, we must regard the whole works—the source, box, gadgets and the cat—as the "quantum mechanical system". This system has two states corresponding to the two alternatives possible:

$|\text{I}\rangle$ – The radioactive atom has not emitted any particle; the hammer has not fallen; the bottle is not broken; and the cat is alive.

$|\text{II}\rangle$ – The atom has emitted a particle; the detector has registered a particle; the hammer has fallen; the bottle is broken; and the cat is dead.

According to QM, at the end of one hour, the wave function of the system (with the lid closed) must be represented as a superposition of the wave functions of the two states, i.e., we must write

$$|\Psi\rangle = \{|\text{I}\rangle + |\text{II}\rangle\} \qquad (3.1)$$

I should really be more careful and put in a factor $(1/\sqrt{2})$ for purposes of normalisation but such details are not important at present. Forgetting all objects except the cat, this can be written in the simpler form

$$|\Psi\rangle = |\text{Alive}\rangle + |\text{Dead}\rangle. \qquad (3.2)$$

This is what QM says, provided the lid is not opened.

As Schroedinger puts it,

> The ψ-function [Schroedinger always referred to the wave function as the ψ-function] of the entire system would express this by having in it the living and the dead cat (pardon the expression) mixed or smeared out in equal parts.

Schroedinger argued that this whole thing is ridiculous. Either the cat is alive or it is dead, even if we do not look into the box ourselves. But $|\psi\rangle$ in (3.2) gives only a "blurred" vision of reality, i.e., a cat that is both alive and dead. So the interpretation given to $|\psi\rangle$ is not correct.

Notice the clever manner in which Schroedinger argues. If we deal with a single radioactive atom whose state we do not know about, we would write

$$\{|\text{ Not decayed }\rangle + |\text{Decayed}\rangle\}. \qquad (3.3)$$

Everyone would agree with this statement and no eyebrows would be raised because we are dealing with a microscopic system. We have a superposition of states because of an indeterminacy in the state of the atom — we don't know if it has decayed or not.

Now it is true QM was discovered in the process of trying to explain physics at the microscopic level. However, there is nothing that says that the theory is applicable *only* at the microscopic level. What Schroedinger decided to do was to apply it at the macroscopic level, i.e., to a cat. In the case of an atom we are not unduly bothered when we write a wave function as in (3.3). This is mainly because we cannot ever see an individual atom and are not able to directly relate to it. On the other hand, in the case of the cat we have a distinct feel for what is going on, and it does seem odd to say that the cat is a state which is a superposition of a live state and a dead state. But then we cannot have a theory in which there is one rule for a cat and another for an atom. Since the interpretation of the wave function in the case of the cat does not make any sense, one must conclude that it does not hold in the case of the atom either — this in brief was Schroedinger's way of looking at the whole thing. As he himself put it:

> It is typical of these cases that an indeterminacy originally restricted to the atomic domain becomes transformed into macroscopic indeterminacy, which can then be resolved by direct observation. That

26   What is reality?

prevents us from so naively accepting as valid a "blurred model" for representing reality. In itself it would not embody anything unclear or contradictory. There is a difference between a shaky or out-of-focus photograph and snapshot of clouds and fog banks.

What do we do now? There are aspects of this question that we shall examine in a later chapter but for the moment I shall restrict myself to telling you about the CI lobby's reaction to Schroedinger. Their view is simple: "The only relevant thing here is the atom and its state because the atom alone forms the quantum system. The rest belong to the *apparatus* which, remember, is classical. Thus the cat is a part of the measuring apparatus in this experiment, and we have no business writing a wave function in the form (3.2) which is permitted only in QM. In other words, it is illegal (at least in this case), to translate a *microscopic uncertainty* (associated with the state of the atom) to a *macroscopic uncertainty* (associated with the state of the cat). Schroedinger is therefore wrong in trying to read meaning into the wave function, the way he is doing."

One could push this "explaining away" even further by adding what I might call a "thermodynamic twist". What really happens when we observe? Basically, we are trying to obtain information. But it is not possible to obtain information without increasing entropy (for a discussion of entropy etc., see the companion volume *A Hot Story*). This increase of entropy is linked with irreversibility and the arrow of time. The point simply is that an observation made on the quantum system transfers the information irreversibly to the macro world, in the process increasing entropy—the system cannot thereafter slip back into the "quantum never-never land". In the case of the cat, it is a macroscopic object. Its state refers to the macro world, and whether we open the lid or not the information about its state exists in the macro world. The generation of this information is accompanied by an increase of entropy. The fate of the cat is decided whether we open the lid or not and QM cannot be extended here.

There is also an ensemble interpretation to the cat paradox. According to this, the wave function (3.1) represents an ensemble of systems, i.e., a whole bunch of boxes each containing its own deadly contraption and a cat. In half the boxes the atom has not decayed and therefore the cat is alive whereas in the other half, the atom has decayed and naturally the cat is dead. When the experimenter opens the box, all he does is to discover which of the two subensembles (I or II), his particular case belongs to.

I don't believe Schroedinger was ever converted by all these arguments but the world at large went along with the view that there was no paradox at all, which is why the cat paradox is seldom mentioned nowadays in the scientific literature. Most people think the issue is "settled". Is it really?

Many questions can be raised to the CI supporters: "OK, the cat is a part of the apparatus and I can't write a wave function as in (3.2). But somewhere between the atom and the cat is the boundary between the observed quantum system and the classical apparatus. Please tell me where it is." No satisfactory answer. Another question: "Please tell me what equations of QM describe the collapse of the wave function." Complete silence! "Alright, what about the atom itself? It cannot be in a schizophrenic state unable to decide whether it has decayed or not. What does QM have to say about it?"

The CI lobby now becomes quite vocal: "Listen, you've got it all wrong. QM is a probabilistic theory. Such a theory by its very nature *can never say anything about the outcome of ONE single experiment or what one would see in ONE single observation.* When you analyse a coin toss do you ever talk of one toss? Don't you accept a probabilistic forecast? Why can't you accept something similar now? Why can't you accept that it does not make any sense whatsoever to talk of the decay of *ONE* atom or even the fate of *ONE* cat, just as it does not make sense to talk of just one toss of a coin? *Quantum mechanical description is the description of an ensemble or a collection of identically-prepared systems subjected to identical observational tests.*"

The debate does not end—it has been going on and on. One side says for example: "I agree QM works etc., but I *do* want to know what happens in the case of one single system. Suppose I wish to apply QM to the whole universe. How can we talk of ensemble in this case? We have only one universe, don't we?" And the other side replies: "Look, when will you ever learn? You are asking questions on the basis of your *classical* experience. This is a *new* world and the ball game is *different*. You can't ask the usual kind of questions here; it would be stupid to do so."

Despite the shrugging away by the devotees of Bohr, the questions remain. And they are increasingly being raised by big shots themselves. As an example, let me now quote what the well-known theoretical physicist Roger Penrose (who has written a best-seller entitled *The Emperor's New Mind*) says about this business. Penrose describes the evolution of the wave function according to the Schroedinger equation as the **U** (or unitary) process and the collapse or the reduction of the wave function as the **R** (or reduction) process.

28 What is reality?

He says:

> The deterministic process **U** seems to be the part of quantum theory of main concern to working physicists; yet philosophers are more intrigued by the non-deterministic *state-vector reduction* **R** (or, as it is sometimes graphically described: *collapse of the wave function*). Whether we regard **R** as simply a change in the "knowledge" available about a system, or whether we take it (as I do) to be something "real", we are indeed provided with two completely *different* mathematical ways in which the state-vector of a physical system is described as changing with time. For **U** is totally deterministic, whereas **R** is a probabilistic law; **U** maintains quantum complex superposition, but **R** grossly violates it; **U** acts in a continuous way, but **R** is blatantly discontinuous ... *Both* **U** and **R** are needed for all the marvellous agreements that quantum theory has with observational facts ... There is, however, no clear rule, as yet, as to when the probabilistic rule **R** should be invoked, in place of the deterministic **U**. What constitutes "making a measurement"? Why (and when) do squared moduli of amplitudes "become probabilities"? Can the "classical level" be understood quantum-mechanically?

These are deep and puzzling questions. We can bypass them for the moment, but we *have* to come back to them sooner or later. J.S. Bell who will make his appearance in the next chapter, has a view similar to that of Penrose. He remarks:

> So long as we do not know exactly when and how it [wave function reduction] takes over from the Schroedinger equation, we do not have an exact and unambiguous formulation of our most fundamental physical theory.

Figure 3.2 gives a broad overview of the relationship between the **U** and the **R** processes.

You might wonder what your have learnt so far. Essentially the following:

- According to Bohr, reality is created by the act of observation.
- Related to this is the collapse of the wave function.
- Neither Einstein nor Schroedinger liked this collapse business and the related picture of physical reality.
- QM does not describe the collapse of the wave function.

The cat paradox

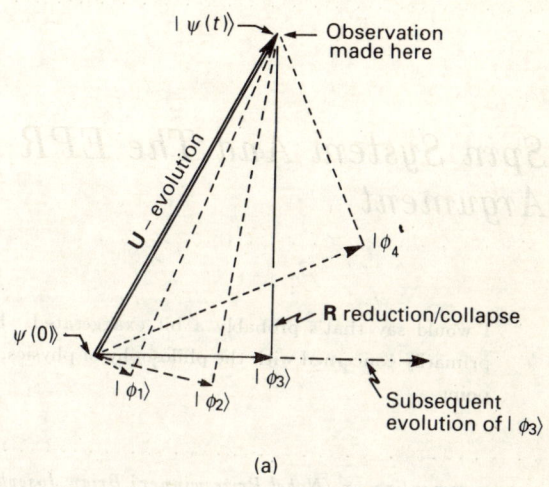

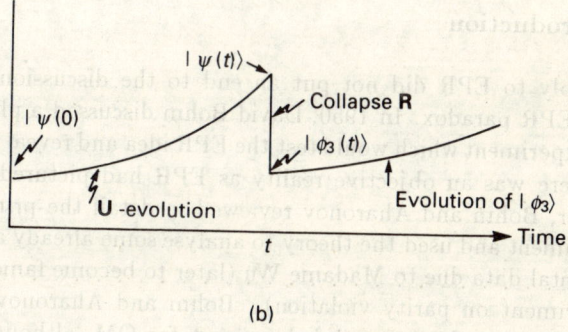

Fig.3.2 An overview of the wave function collapse. To start with, the system evolves from $|\psi(0)\rangle$ by a U-process, i.e., $|\psi(t)\rangle = e^{-iHt/\hbar}|\psi(0)\rangle$ which is just the solution of the time-dependent Schroedinger equation written in symbolic form. At time t a measurement corresponding to observable $\hat{O}$ is made. The state vector then jumps to $|\phi_i\rangle$, one of the eigenvectors of $\hat{O}$. This is illustrated in (a) which shows the various vectors in an appropriate vector space. In this case, the jump is to $|\phi_3\rangle$. Different tries would result in a collapse to the different eigenfunctions of $\hat{O}$. QM has rules for predicting the probabilities. In (b) is shown the time sequence of events (see Fig. 3.15 in Part I). The **U**-process is deterministic whereas the **R**-process is probabilistic.

4 Spin System And The EPR Argument

> I would say that's probably a bit exaggerated. But if you're primarily concerned with the philosophy of physics, I can see his point.
>
> *J.S. Bell,*
> *commenting on (Nobel Prize winner) Brian Josephson's remark that Bell's inequality is the most important recent advance in physics.*

4.1 Introduction

Bohr's reply to EPR did not put an end to the discussion on the so-called EPR paradox. In 1950, David Bohm discussed a physically feasible experiment which would test the EPR idea and reveal whether or not there was an objective reality as EPR had pictured. Eight years later, Bohm and Aharonov reviewed in detail the principle of the experiment and used the theory to analyse some already available experimental data due to Madame Wu (later to become famous with her experiment on parity violation). Bohm and Aharonov argued that the Wu experiment provided support for QM, although in an indirect manner. From that date onwards, there has been a strong interest in the experimental test of the EPR argument. Thanks to Bohm and Aharonov, the debate shifted from dingy rooms filled with pipe-smoking philosophers to labs filled with smart experimenters! In the next chapter, I shall tell you all about experiments and the results obtained.

Upto the time Bohm and Aharonov came up with their proposal, the EPR question was usually discussed in terms of gedanken experiments involving particles, their positions and their momenta. Bohm and Aharonov turned the focus on particle spin which is more conveniently measured. In fact they did one better by drawing attention to the possibility of carrying out such experiments with photons. This was an excellent suggestion because such experiments are more easily done with photons than say with electrons, and indeed most of the experiments performed to date have actually employed photons.

An experiment in physics usually results in a number which is compared with theory to judge whether the theory is correct or not. In this case, clearly QM is a candidate theory. What about the theory, if any, representing the EPR camp? You will recall that EPR's parting shot in their paper was that (i) QM is incomplete as a theory of physics, and (ii) a complete theory was possible which would provide room for objective reality, in the process getting rid of the probabilistic forecasting which is all that QM offers.

4.2 On hidden-variables theories

A few words now about how to construct such a "dream theory". Experience with statistical mechanics suggests that indeterminism enters when there is ignorance about information that is otherwise precise and definite (for a discussion of statistical mechanics, see the companion volume *A Hot Story*). One could likewise argue that probabilities appear in QM because of our ignorance. Of what? Presumably of suitable variables. This might sound like passing the buck but we must concede that the existence of presently unknown variables is a logical possibility.

Suppose we start now by assuming that there exist variables q_1, q_2, q_3,... which at the moment we do not know anything about. It is this ignorance that has led us to develop a probabilistic theory like QM. If somehow we could discover this set $(q_1, q_2,...)$ which at present is *hidden* from us, and we are able to build a theory using this set, we would have a *deterministic* theory which could replace QM. Such a theory is called a *hidden-variables theory*.

The idea that there could exist such a theory is not as stupid as it might sound. Recall Einstein's remarks to Pais about the Moon. And surely we believe that the Universe existed before life appeared on Earth and humans emerged by a long evolutionary process. Objective reality is therefore a strong component of our instinctive feeling and cannot just be wished away that easily, inspite of all that Bohr might say.

So as an alternate to QM, EPR hinted at the hidden-variables theory as a suitable candidate. If you like a political tinge, you might say that QM is a candidate of the Epistemological party and the hidden-variables theory a candidate of the Ontological party! Voting

32 What is reality?

now becomes easier. All one has to do is to calculate the outcome of any proposed experiment according to the two theories, and see which one the experimental results actually favour. More about this later.

4.3 The Bohm–Aharonov proposal

Let us now consider the essentials of the Bohm–Aharonov proposal. Their idea can be understood with the help of Fig. 4.1. A pair of electrons is produced at S under special conditions about which I shall say more shortly. The two electrons 1 and 2 fly off in opposite directions as shown. (For convenience, I might sometimes refer to these electrons as A and B.)

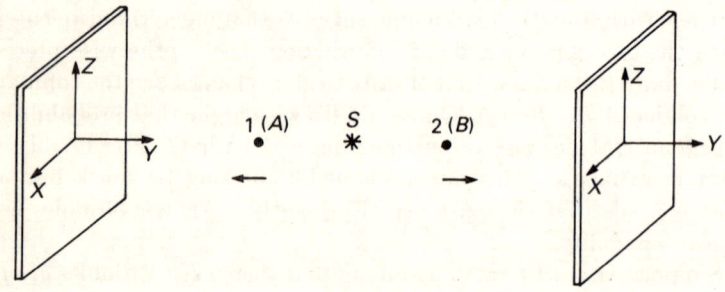

Fig.4.1 EPR experiment on spin states. A pair of particles is produced in the singlet state at S. The two particles are then shot in opposite directions. But they carry with them the spin information of the parent state. This shows up when the spins of the individual particles are measured. The axes show convenient reference frames for the measurement of the spin components.

As you are aware, the electron has a spin angular momentum of $(\hbar/2)$. Let me remind you what this implies. Firstly, if one measures the component of the electron spin angular momentum along any particular direction, one would obtain the value $(\hbar/2)$ or $-(\hbar/2)$. Say the direction chosen is the z-axis. Then s_z can take only the values $(1/2)$ or $-(1/2)$ (in units of $\hbar$). The second point is that once s_z is measured, the other two spin components namely s_x and s_y cannot be given precise values because s_z does not commute with either s_x or with s_y. In other words, at a given time, only one of the three spin components can have physical reality in QM. Before the actual measurement, all three components have potentialities but once the measurement is made, one component alone becomes a reality; the other two fluctuate randomly and become irrelevant as far as reality is concerned (in QM that is).

A few words now about the special state in which the electron pair is created. Given two electrons, their individual spins can combine to produce a total spin of either 1 or 0. The former combination is called the *triplet* state and the latter the *singlet* state. Our interest is in the singlet state. If while the pair is in this state we determine the z-component of the spin of electron 1 and find s_z to be $+(1/2)$, then s_z for electron 2 MUST equal $-(1/2)$. Conversely, if electron 1 has spin of $-(1/2)$, then electron 2 must have a spin of $+(1/2)$. I remind you again that once s_z for electron 1 is measured, the components s_x and s_y become irrelevant for that electron. This is standard QM wisdom and must be constantly borne in mind.

The points I have just made are so important that you would forgive me if I repeat all that over again in a slightly different form. In the singlet state, the spin components obey the following equations:

$$s_{Ax} + s_{Bx} = 0 \tag{4.1}$$

$$s_{Ay} + s_{By} = 0 \tag{4.2}$$

$$s_{Az} + s_{Bz} = 0. \tag{4.3}$$

Using these equations, we can predict the component of particle B (or 2), if the corresponding component of particle A (or 1) is measured. For example, if a measurement shows 1 to have s_z equal to $+(1/2)$, then we know from (4.3) that for particle 2, $s_z = -(1/2)$. We could just as well have chosen to measure s_x (or s_y) for particle 1 and inferred the value of s_x (or s_y) for particle 2. But QM forbids us from knowing at a time the value of more than one component. The choice before us is: Measure s_{Ax} and infer s_{Bx}, or measure s_{Ay} and infer s_{By}, or measure s_{Az} and infer s_{Bz}. One could of course perform the measurements on particle 2 and draw conclusions about particle 1 — that is our choice.

Next let $\Psi_+(1)$, $\Psi_-(1)$, $\Psi_+(2)$, and $\Psi_-(2)$ denote the one-particle wave functions for the particles 1 and 2 in the states $+(1/2)$ and $-(1/2)$ respectively. Form now the combinations:

$$\begin{aligned}\Psi_{+-} &= \Psi_+(1) \cdot \Psi_-(2), \\ \Psi_{-+} &= \Psi_-(1) \cdot \Psi_+(2).\end{aligned} \tag{4.4}$$

Two other combinations namely, Ψ_{++} and Ψ_{--} can also be similarly formed but we shall restrict to those in (4.4) as they alone pertain to the singlet state which is of current interest. Using the kit in (4.4) let us now form the *two-particle* wave function

$$\Phi(1,2) = (1/\sqrt{2})\{\Psi_+(1)\Psi_-(2) - \Psi_-(1)\Psi_+(2)\}. \qquad (4.5)$$

This describes the pair of electrons in the singlet state. It is what is called a *pure* state. Observe that Ψ_{-+} is added to Ψ_{+-} with a definite phase (which corresponds to a numerical value of -1). I must emphasise that (4.5) represents the wave function *before* any apparatus has interacted with any one of the two particles. At this stage of the game we cannot say anything about any of the spin components of the two particles, other than that various potentialities exist.

The two-particle system is now allowed to interact with an apparatus. This could, for example, be one designed to measure the z-component of particle 1. Once the apparatus interacts with one of the two particles in the system, all the potentialities drop out of the picture except of course the relevant one which becomes a concrete reality. In the wave function language, Φ collapses to either Ψ_{+-} or to Ψ_{-+}, depending on whether upon measurement particle 1 — we assume that it is particle 1 which is being observed — has a spin of $+(1/2)$ or $-(1/2)$. Besides causing a collapse of the wave function, the apparatus also disturbs the total spin angular momentum of the two-particle system, i.e., the total spin is no longer equal to zero. As Bohm puts it:

> After the measurement is over, the system has been transformed from one that had a definite combined angular momentum and an indefinite value of s_z for each particle to one which has a definite value of s_z for each particle, but an indefinite combined angular momentum. Moreover, the precise value of s_z which will be obtained for each particle is not related deterministically to the state of the system before the measurement [i.e., to $\Phi(1,2)$] but only statistically ... The mathematical description provided by the wave function [$\Phi(1, 2)$] is certainly not in one-to-one correspondence with the actual behaviour of the system under description, but only in a statistical correspondence.

In short, although the wave function does contain the most complete possible description of the system as allowed by QM, this description is incapable of precisely defining the reality that would ensue once a measurement is made. All this is simply a long-winded way of saying that QM has no deterministic prescription for predicting the outcome of a wave-function collapse. (See also Box 4.1.)

> **Box 4.1** If you are technically minded, you might worry what would happen if the spin components are measured in an arbitrary direction $\hat{n}$ as in figure (a). It turns out that since the singlet state is spherically symmetric, we can write the wave function as $\Phi(1,2) = 1/\sqrt{2}[\psi_{\hat{n}}^+(1)\psi_{\hat{n}}^-(2) - \psi_{\hat{n}}^-(1)\psi_{\hat{n}}^+(2)]$. I think the notation should be clear. Basically, the spin-up and spin-down functions for the two particles have been defined with respect to the new direction $\hat{n}$. Ψ_{+-} and Ψ_{-+} are now defined in terms of $\psi_{\hat{n}}^+(1)$ etc. The rest of the arguments in the text go through without any hitch.
>
> (a)

OK, so we have these two particles existing as a pair in a singlet state described by $\Phi(1,2)$ given in (4.5). What exactly was the proposal of Bohm and Aharonov? Briefly, what they said was that (4.5) (which was OK as far as QM was concerned), was not consistent with EPR's concept of reality. Whereas according to standard QM the outcome of a measurement is determined by an ensemble whose members are described by (4.5), the EPR view of reality demands that the ensemble is a *mixture* of states of the type $\Psi_+(1)\Psi_-(2)$ and $\Psi_-(1)\Psi_+(2)$. Bohm and Aharonov showed that in certain kinds of experiments, there was a *measurable difference* between the two points of view; that is why their analysis is important—see also Box 4.2.

36 What is reality?

> **Box 4.2** This box also is for the technically minded who want to dig a little deeper. Consider a two-particle system $\epsilon = (\alpha, \beta)$. Let $(|\Psi_1\rangle, |\Psi_2\rangle ...) = \{|\Psi_i\rangle\}$ denote a complete set of state vectors describing α. Similarly, let $\{|\Phi_j\rangle\}$ be a complete set appropriate to β. Consider now the combination
>
> $$|\eta\rangle = \sum c_{ij}|\Psi_i\rangle|\Phi_j\rangle \qquad (1)$$
>
> where c_{ij} are complex coefficients. Of course they are subject to the normalisation condition
>
> $$\sum_i \sum_j |c_{ij}|^2 = 1. \qquad (2)$$
>
> Suppose $\widehat{A}$ and $\widehat{B}$ are Hermitian operators such that
>
> $$\widehat{A}|\Psi_i\rangle = a_i|\Psi_i\rangle \cdot \widehat{B}|\Phi_j\rangle = b_j|\Phi_j\rangle. \qquad (3)$$
>
> Choose c_{ij} such that
>
> $$c_{ij} = a_i b_j. \qquad (4)$$
>
> $|\eta\rangle$ with c_{ij} as in (4) is one possible description of the system (α, β). States of this kind are sometimes said to be of the *first* type. Essentially they are mixtures of the product wave functions $|\Psi_i\rangle|\Phi_j\rangle$. It is possible to have states $|\eta\rangle$ wherein c_{ij} cannot be written as a product as in (4). One example of such a state is in (4.5). States of the latter kind are said to be of the *second* type. In contrast to the first type, in a state of second type neither α nor β are in well-defined quantum state. This is important. For example, in the case of (4.5), we cannot say that 1 is in the state $s_z = +(1/2)$ and 2 in the state $s_z = -(1/2)$. States of the second kind are *pure*.
>
> In the case of some experiments, the outcome can equally well be explained by starting with ensembles of pure states or of mixtures. But the Bohm–Aharonov experiment is different. In this experiment there is a *testable* difference between the two cases. And the question boils down to: Which of the two is actually favoured by Nature?

4.4 More about the alternative to QM

What I have said so far is standard QM stuff. Let us now zoom in on EPR's objection. Basically they take issue with potentialities, wave function collapse that is supposed to occur when an apparatus interacts, and the sudden emergence of reality from what until then was a vague bunch of potentialities. They would rather prefer a deterministic theory (a hidden-variables theory perhaps) which makes room for objective reality and locality, at the same time doing away with probabilistic forecasts, wave function collapse, etc. More about this soon.

So far, no one has succeeded in coming up with a hidden-variables theory that would meet the bill. Does that mean there is a stalemate? Not quite. In a very important piece of work, J.S. Bell showed how to calculate the outcome of a EPR-type experiment in a dream deterministic theory, without even knowing the details of the theory. Further, under certain conditions, this dream theory makes very different predictions from standard QM. Careful experiments should therefore be able to reveal whether such a theory at all *exists*, even though we may not know anything about it at the present time. No wonder Bell's work created a great stir. I am sure you would find it difficult to believe that one can predict the outcome of a theory without even knowing what the theory is like; so let me first tell you something about Bell's work.

To start with, we have to be clear about what it is that we are comparing. And the answer is that we are comparing standard QM with a theory founded upon the following three assumptions:

1. Conclusions can be drawn from observations on the basis of logical deductions.
2. Observed phenomena are caused by some reality which exists whether we observe the phenomena or not.
3. Information cannot be transmitted with a speed greater than that of light.

We shall refer to these three assumptions as: *Induction, Realism* and *Local causality* (also sometimes referred to as separability or Einstein separability). Of these three, I would like to say something more about the last by quoting the physicist Heinz Pagels. He observes:

> The basic idea of local causality is the following: Events far away cannot directly and instantaneously influence objects here. If a fire breaks out a hundred miles away there is no way it can directly influence you. A second after the fire breaks out a friend may telephone you and tell you about the fire — but that is ordinary causality. Information about the fire has been transmitted by an electromagnetic signal from your friend to you.

38 What is reality?

Causality is a principle which all physicists accept, and Einstein swore by it. In their paper, EPR declared that either QM must admit it is incomplete or be ready to violate local causality. Since EPR simply could not dream of the latter, they pronounced the judgement that QM is an incomplete theory. Incidentally, a crisp way of stating the condition of local causality is that there is no action-at-a-distance.

4.5 Local causality and the EPR experiment

Einstein attached much importance to local causality. It is therefore useful to spend a minute on the EPR experiment as performed on a pair of electrons in the singlet state. The electrons are well separated — say they are 1000 km apart. What would EPR do and how would they argue? They would say:

1. Measure the z-component of the spin of particle A. Say it is $+(1/2)$. Using (4.3), the value of s_z of B is deduced to be— $(1/2)$.
2. Using the (EPR) criterion for reality, s_{Bz} can now be assigned a reality status.
3. Next measure s_{Ax}. Using (4.1) deduce s_{Bx}.
4. Again, using the reality criterion, assign a reality status to s_{Bx}.
5. In effect, by observations on A alone, we have been able to determine s_{Bz} and s_{Bx}. Therefore these two quantities have simultaneous reality.

This is EPR's point of view. The QM supporters will of course not buy this line of argument. They would say to EPR: "Once you observe s_{Az}, the system is no longer in the singlet state. That state is instantly destroyed. Therefore, after your second measurement, i.e., of s_{Ax}, you have no right to use (4.1)." EPR grin and reply, "Got you! How can a measurement on A instantaneously convey to B that A and B are no longer in partnership as a singlet pair? Would this not imply sending a signal with a speed greater than c? Does that not amount to violating local causality?"

Most of us would dismiss EPR's objections. Indeed, experiments which I shall describe in the next chapter come to our support. And yet we should not overlook how much QM goes against our gut feeling. Consider the following "twin paradox" invented by a philosopher to highlight this point. There is a pair of identical twins separated at birth (this often happens in our movies!). One is brought up in Delhi and the other in Madras. Not only do the twins look alike etc., but their life styles also are quite similar. Suppose the twin in Madras is sitting on a chair. His counterpart in Delhi would also be doing the same at that very instant. Imagine we now push the fellow in Madras so that he falls down from his chair. No surprise in that. But what would you say if the fellow in Delhi also falls down from his chair at that very instant? This is precisely what QM implies, and it is this eerie, nonlocal influence that binds partners because these influences act instantaneously and do so even across the entire Universe! Space seems irrelevant in such matters!! Seen this way, QM does defy commonsense.

I am elaborating this point because later you will see that the hidden-variables theory is itself in trouble on this score. This of course is something that EPR did not bargain for!

4.6 Bertlmann's socks

For a moment let us forget EPR and ask : Why all this fuss about the local causality business. The point is actually about instantaneous correlation between distant events. However, we little realise that even in ordinary life there can be such correlations, without upsetting local causality and all that. Bell gives a humourous example.

There is a gentleman called Dr. Bertlmann who is a bit of a funny character and likes to wear socks of different colours! Which colour he will have on a given foot on a given day is quite unpredictable. But when you see that the first sock is pink, you can be certain that the second sock would NOT be pink. You know this without even looking at the second sock because you happen to know Dr. Bertlmann's habits. There is no instantaneous transmission of information from one leg to another etc. In short, there is no mystery. Is not the correlation of spins something like this? Why should one be particularly bothered about it as EPR seem to be?

You would have noticed that in the case of the socks, there is first an observation (of the colour of the sock on one leg) and then a logical deduction (that the colour of the sock on the other leg is different). Let us follow up this correlation and deduction business a little further.

4.7 Correlations in a psychological test

This time we deal with a psychologist. This chap has come up with a test which will reveal whether people have fear of death or not (you know psychologists do this kind of thing all the time). He administers this test to a large number of people and finds that his subjects either pass the test (which means that they have no fear) or fail in the test (which means that they are susceptible to fear). But the psychologist does not know what distinguishes the two sets of people, i.e., why some people pass and others fail.

The psychologist now tries a different strategy. He selects as his subjects, only married couples. But as before every person is given the test separately. After all the tests are completed, he pores over the data. Considered individually, the results merely show as before that some people pass the test while others fail. However, when the data is analysed in terms of couples, he finds the interesting result that when the husband passes, so does the wife. Likewise, when one fails the other also does. In other words, he finds a clear correlation between the results for the two members of a couple.

He now wonders: "Could it be that when say the husband comes out after taking the test he secretly tells his wife how to answer the questions? In that case, it would be no wonder that the husband and the wife fare in the same way." He then rejects the entire data, and to prevent such conspiracies he makes sure that the husbands and the wives are separated from each other right at the beginning, and are not allowed to communicate with each other in *any possible way* till all the tests are over. This time also he finds as before that the behaviour of the two partners is always the same. The psychologist now has a choice: Either he can dismiss it all as a freak and remarkable coincidence (which would be like getting heads all the time when one repeatedly tosses a coin) or speculate that there is some definite reason why the husband and the wife respond to the test in the same way. This second statement would be a natural and a logical inference from the results of the tests.

I brought in the psychologist mainly to impress upon you that even in this example, we make use of the three assumptions mentioned earlier. We have the idea of realism (people have definite attitudes even before they are actually tested), the idea of separability (the husbands and the wives are separated and prevented from communicating with each other) and the concept of induction (which I just discussed). The alternate to QM is visualised as a theory based on the same three assumptions. Such a theory is called a *local-realistic theory* (LRT) and as you will soon see, a LRT implies hidden variables in practice. I shall therefore regard LRT as synonymous with a hidden-variables theory.

The question before us now is whether in principle a suitable LRT exists which would satisfy EPR, i.e., a LRT which not only predicts what classical theory fails to, but goes further than QM in not having QM's "unsatisfactory" features. EPR believed such a theory could be constructed. The obvious way to check this out is to see what a LRT would predict as compared to QM, and let the actual experiment decide. Before I discuss these delicate matters, maybe it would be good to introduce you to a simpler and amusing version of the EPR-type experiment as proposed by David Mermin. The idea is to show how hidden variables enter into the picture.

4.8 Mermin's imitation

We start with Mermin's experimental arrangement which is shown in Fig.4.2. S is a source which fires pairs of particles, one each towards

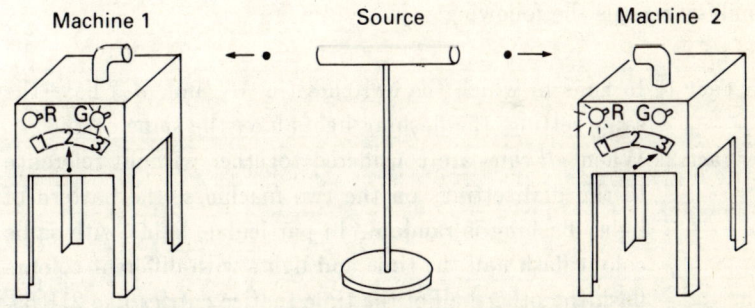

Fig.4.2 Mermin's imitation of the Bohm–Aharonov version of the EPR experiment. Particles are shot in pairs in opposite directions. The horns catch them and direct them to machines M_1 and M_2. Each machine has three settings and two colour bulbs. An example of the record of observations is shown in Fig.4.3.

the two machines M_1 and M_2. The machines register by flashing a bulb. There are two bulbs, one red (R) and the other green (G); one of these two will flash when the particle reaches a machine. Each machine has a dial with three settings 1, 2, and 3. Every firing by S is an experimental run. In every run, the dials on the two machines are set at random. The experiment thus consists in steadily firing from S and recording which light flashes in the two machines, along of course with the machine settings. The log book would thus have entries as below.

M_1 setting	M_2 setting	Bulb flashing in M_1	Bulb flashing in M_2
1	2	R	G
3	3	R	R
2	3	G	R
3	1	G	G
1	3	G	R
2	1	R	R
1	1	R	R
2	2	G	G

For convenience, I shall abbreviate the records as 12RG, 33RR, 23GR etc. Let us suppose a long record of results has been assembled—see Fig.4.3—and the data is analysed. I shall also assume that the data analysis reveals the following:

Fact 1: In runs in which the switches (in M_1 and M_2) have the same setting, the flashing lights have the same colour.
Fact 2: When *all* runs are considered together without reference to the dial settings on the two machines, the pattern of lights flashing is random. In particular, lights with same colour flash half the time and lights with different colours flash the other half of the time (notice entries like 21RR).

Remember this is what the experiments are supposed to give and that so far, theory has not entered the picture. Theory enters when one tries to understand the two facts mentioned above.

I am sure you would have spotted all the analogies between Mermin's experiment and the original Bohm–Aharonov proposal. For your convenience, let me list them below:

	Mermin	Bohm–Aharonov
1.	Particle	Electron
2.	Machine	Spin detector
3.	Machine settings 1,2,3	Directions x, y, z
4.	R, G	$(+\,1/2)$, $(-\,1/2)$

Spin system and the EPR argument   43

13RG	13GR	12RG	22RR	22GG	11GG
31RR	22GG	33GG	12GR	22RR	21RG
33GG	11RR	21GR	32GR	13RR	11RR
33RR	33GG	13GR	22RR	21GG	12RG
12GR	31GR	23GR	12GG	23GR	23GR
33GG	12GG	22RR	33RR	22GG	32GR
21GR	21GR	11RR	11RR	22GG	21GG
21RR	33GG	21GR	23GG	31GG	21RG
22RR	21RR	21RR	23GG	13GR	13RG
33GG	12GR	23GG	33RR	21GR	13RG
11GG	22RR	32GR	23GR	33RR	13RG
23RR	13RG	33RR	21GG	23RR	13GR
32GR	12RG	33GG	13GR	22RR	23RG
12GR	23GG	33GG	33GG	12RR	22GG
12RG	11GG	23GR	11GG	23RG	11RR
11GG	13RG	21GR	12RR	23RG	31RG
31RG	21RG	12RR	12GG	32GR	23RR
12RG	33RR	32GR	31GG	31RG	23RG
13GR	32GR	32GR	32RG	22GG	11RR
22GG	32GG	33GG	21GR	11GG	32RG
12RG	33GG	31RG	22GG	11GG	32GR
12GR	21RR	13RR	22RR	21RG	13GG
22GG	12RG	13RG	13RR	11RR	23GR
23GR	22GG	32RG	21GG	12RG	32GR
33RR	11GG	31GR	23GR	22GR	22RR
33GG	23GR	23RR			
31RG	22RR	33RR			
31RR	11GG	13RG			
33RR	32GR	11GG			
32RG	13RG	31GR			
31RG	13GR	31RG			
11RR	23GG	13GR			
23GR	33RR	23RG			
12GG	31GR	31GG			
11GG	13RG	23RG			
13RG	23RR	21RR			
31RG	12GR	23RG			
23GR	31RG	11GG			
31GR	32RG	22GG			
23RG	21GR	11GG			

Fig.4.3 Typical data produced by the apparatus of Fig.4.2. The notation and the significance of the data are explained in the text.

44 What is reality?

4.9 An attempt to explain the results of Mermin's experiment

Let us now try a model which mimics the hidden-variables kind of theory and attempts to explain the results of Mermin's experiments. After Mermin, we shall suppose that in each machine there is a target disc divided into eight sectors which are labelled as in Fig.4.4. There

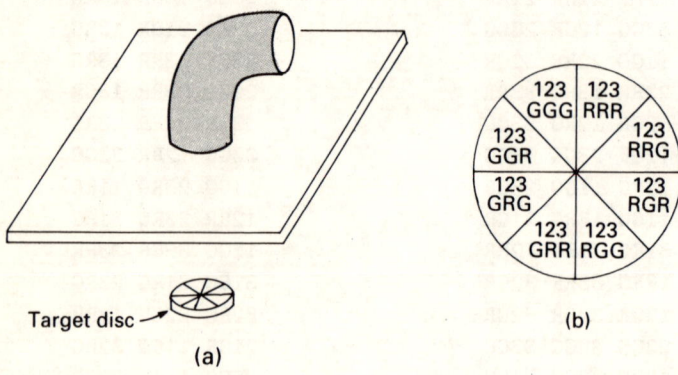

Fig.4.4 Hidden-variables interpretation of the apparently random data of Fig.4.3. In this picture, inside each machine is a target disc with 8 sectors as in (a). The details of the sectors are shown in (b). The particles are fired by the source to hit a particular sector; both particles are coded to hit the same sectors in their respective machines. The sector coding is randomly altered from run to run. How all this helps to predict the outcome is discussed in the text.

is a complicated circuit which links these sectors to the dial settings and to the lamps. Thanks to all these linkages, if the particle arriving at a machine hits say the sector RGR, then R will glow if the dial is set in position 1, G will glow for setting 2 and R will flash for setting 3. The same is the case when a hit is made in some other sector. Obviously, if the sector is either RRR or GGG, then the same bulb will flash irrespective of the dial setting. If we did not know anything about these sectors, we would think the machine works in a probabilistic way. But if the particles leave the source S with instructions about the sector they are to hit (particles in a given pair are supposed to hit identical sectors, i.e., both RGR or both GGR etc.) then the outcome is completely predictable and there is nothing probabilistic about the whole business. The instruction sets, the sectors, the wiring, etc., are all hidden from us.

Even though we don't know many of the details, we can as it is predict what such a model would yield as results. In particular, we can check if the model predicts the experimentally observed fact number 2. Let us say that particles have arrived at the two machines M_1 and M_2 with instructions to hit the sector RRG. This means that both machines would flash R for the (joint) settings 11, 12, 21 and 22. The two machines would flash G for the setting 33. The machines would flash different colours for the settings 13, 31, 23 and 32. Thus for five settings out of a total possible of nine, the same colours would flash. This score of 5/9 applies to six of the eight sectors, i.e., to all except sector RRR and sector GGG. Everything taken together, we can say that same colour lights would flash *at least* 5/9 of the time, if not slightly more. Compare this with the experimental fact that the same colour lights flash only 1/2 of the time. Thus in this contrived example, the hidden-variables theory does not agree with experimental data. With this model example in mind, we are now ready to take in more weighty stuff!

4.10 Enter Bell

As I told you earlier, the EPR paper was published in 1935. For the next thirty years there was only a debate with people arguing in all sorts of ways and uttering a lot of mumbo jumbo. Things changed in a sharp manner in 1965 when John Stuart Bell entered the picture.

Bell took the Bohm–Aharonov gedanken experiment seriously, and decided to pit QM against the hidden-variables or the local-realistic type of theory. He considered a pair of spin one-half particles formed somehow in the singlet spin state and moving freely in opposite directions. Using suitable pieces of apparatus, select spin components of the two particles are then measured. For particle 1 it could be the z-component and for particle 2 it could be the x-component. Now we must be a bit careful about specifying the components, and this calls for introducing suitable reference vectors. Before we do that first let us set up a universal set of axes as in Fig. 4.5. With reference to these axes I can say I am measuring the z-component of the spin of 1 and the x-component of the spin of 2 and so on. But we wish to be a bit more general and so introduce two unit vectors **a** and **b** as shown. We now decide that we will always be interested in the component of

46 What is reality?

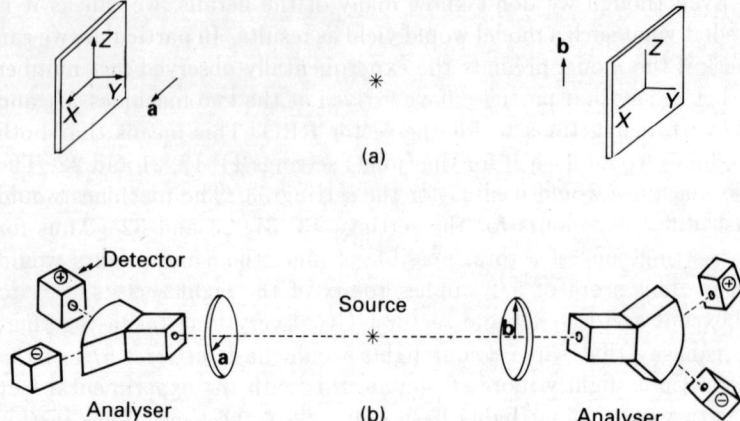

Fig.4.5 Schema of arrangement envisaged by Bell. (a)shows the source, the universal set of axes and the special directions **a** and **b**. In each observation station, there is a pair of detectors waiting to pick up particles which are up or down with respect to the reference direction for that station. This is shown in (b). It is from these detector counts that the quantity $E(\mathbf{a},\mathbf{b})$ mentioned in section 4.11 is deduced.

spin of particle 1 along direction **a** and the component of spin of particle 2 along **b**. According to QM, along a given direction, spin angular momentum can take the values $+(\hbar/2)$ and $-(\hbar/2)$. Let us take the unit of spin as $(\hbar/2)$. Then the possible values for the spin components are $+1$ and -1. QM says that the product of the values of the spin components is given by $-\mathbf{a} \cdot \mathbf{b}$; more about this soon.

Bell showed that a hidden-variables theory (based on the three assumptions mentioned earlier) gives a different result compared to QM, though in the special case of **a** parallel to **b** it gives the same result. Bell's conclusion can be stated as follows:

> No hidden-variables theory based on the assumptions of reality, inductive logic and local causality or separability can agree with *all* the predictions of quantum mechanics. In other words, under certain conditions its predictions for the outcome of an EPR-type experiment on the spin components of particles in the singlet state are *different* from those of QM. Incidentally, this also implies that we *cannot embed* QM in a LRT. A careful experiment can therefore decide whether QM is correct or the hidden-variables/local-reality theory.

4.11 Bell's inequality

The great advance made by Bell was that he focussed attention on making predictions which could be compared with experiments. In his paper, Bell derived a particular relation called *Bell's inequality*. No reference to his work would be complete without some mention of this famous result.

Unfortunately, I have not been able to get much biographical material about this remarkable Irish scientist. John Stuart Bell was born in 1928 and died in 1990. Prof. R. Rajaraman of the Centre for Theoretical Studies, Indian Institute of Science writes:

> His path-breaking work on the foundations of quantum theory, acclaimed of course by the physics community, had also made him famous in a large world—a celebrity status that he handled with quiet dignity and gentle amusement. As a person, he was kind and softspoken, yet commanded much respect, sometimes bordering on awe.

Let us now get back to Bell's inequality. Consider Fig.4.6(a). We have four directions $\mathbf{a}$, $\mathbf{a}'$, $\mathbf{b}$, $\mathbf{b}'$. Measurements are made at the two stations, choosing the pairs of reference directions $(\mathbf{a}, \mathbf{b})$, $(\mathbf{a}, \mathbf{b}')$, $(\mathbf{a}', \mathbf{b})$ and $(\mathbf{a}', \mathbf{b}')$. For each pair, a series of measurements are made and a certain correlation function is evaluated.

Suppose the pair $(\mathbf{a}, \mathbf{b})$ is chosen. Say observations are made on N pairs of particles (α, β). On particle α observations are made with reference to direction $\mathbf{a}$ and on β they are made with reference to direction $\mathbf{b}$. Each measurement (be it on α or on β) yields either $+1$ or -1. The N measurement yield the sets $\{A_1, A_2, \ldots A_N\}$ and $\{B_1, B_2, \ldots B_N\}$ where $A_i = +1$ or -1 and $B_i = +1$ or -1. The correlation function $E(\mathbf{a}, \mathbf{b})$ of the results A_i and B_i is defined as the average of the product $A_i B_i$, i.e.,

$$E(\mathbf{a}, \mathbf{b}) = \frac{1}{N} \sum_{i=1}^{N} A_i B_i. \tag{4.6}$$

For two spin $(1/2)$ particles in the singlet state (4.5),

$$E(\mathbf{a}, \mathbf{b}) = -\mathbf{a} \cdot \mathbf{b}. \tag{4.7}$$

Define now

$$\Delta \equiv |E(\mathbf{a}, \mathbf{b}) - E(\mathbf{a}, \mathbf{b}')| + |E(\mathbf{a}', \mathbf{b}) + E(\mathbf{a}', \mathbf{b}')|. \tag{4.8}$$

48 What is reality?

According to local realistic theory, the maximum value Δ can take is 2 whereas based on (4.7), i.e., a state function of the second type, the maximum value for Δ is $2\sqrt{2}$. These maximum values occur for the choice of reference directions shown in Fig. 4.6(b). Bell's inequality is stated in the form:

$$\Delta \leq 2 \qquad (4.9)$$

Observe that unlike in EPR's original proposal, Bell is not restricting himself to an individual event; rather, he is studying the statistical correlation of N events. For certain choice of the directions of the four vectors $\mathbf{a}$, $\mathbf{a'}$, $\mathbf{b}$, $\mathbf{b'}$, the value of Δ derived from QM is the same as that predicted by LRT. But for certain other choices, e.g., that in Fig.4.6(b), Bell's inequality is violated by QM.

The strategy is now clear. (i) Set up a Bohm–Aharonov type experiment. (ii) Measure Δ. Do it particularly for the case where LRT and QM make different predictions. (iii) Compare experimental results with theoretical predictions, and select the winner. The flow chart in Fig.4.7 says the same thing, together with a few additional points to which I shall come later.

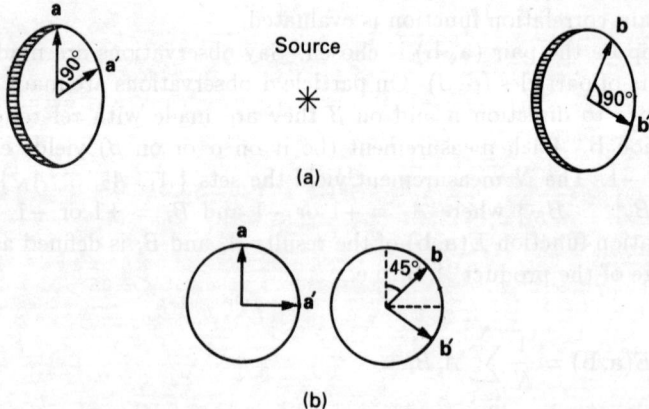

Fig.4.6 (a) Polarisation directions pertinent to (4.8). (b) The special case when Δ in (4.8) is a maximum.

It is pertinent to stress once again some of the important features of the experimental situation considered by Bell. True he starts like EPR (or rather like Bohm–Aharonov). But he generalises in two

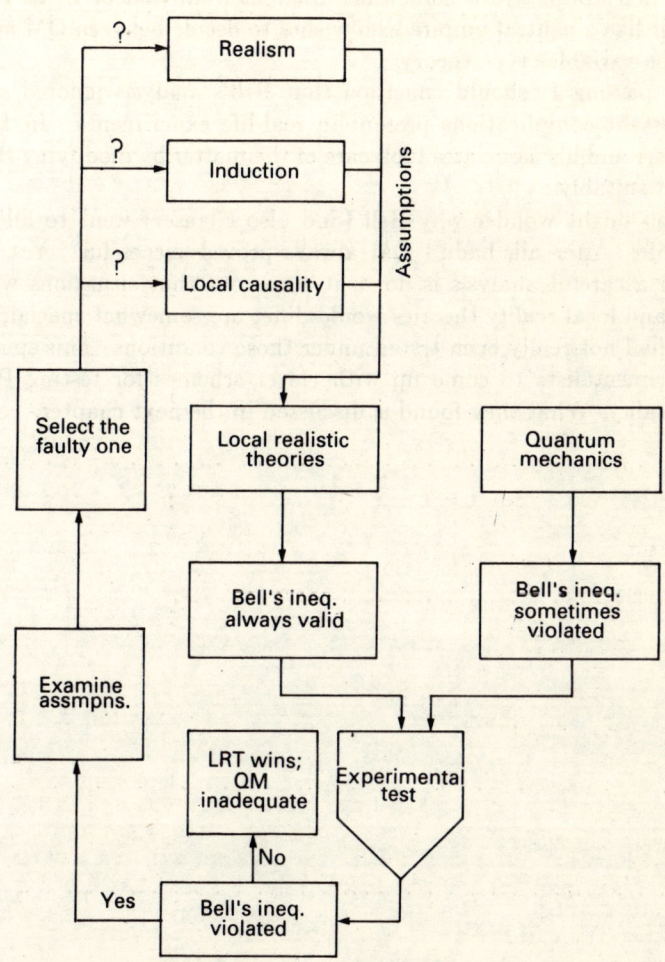

Fig.4.7 Flow sheet for the experimental test of Bell's inequality. Basically there are two candidates. The local-realistic theory must respect the inequality. If experiments violate the inequality, then LRT is in trouble. The issue then boils down to which of the three assumptions underlying LRT must be modified.

respects. Firstly, he does not rigidly restrict the directions of polarisation; in other words, he allows directions **a** and **b** to vary and not merely be parallel. Secondly, he is not looking at individual events but at the averages of a collection of them. We must also remember that Bell's objective is somewhat different from that of EPR. He is rather like a neutral umpire who wishes to decide between QM and a hidden-variables type theory.

In passing I should mention that Bell's analysis ignored some important complications present in real-life experiments. In 1969, Clauser and his associates took care of the matter by modifying Bell's result suitably.

You might wonder why Bell (and also Clauser) went to all this trouble. After all, hadn't QM always proved successful? Yes, but when a careful analysis is done, it turns out that situations where QM and local-reality theories would differ are somewhat special; and QM had not really been tested under those conditions. This spurred experimentalists to come up with clever schemes for testing Bell's inequality. What they found is discussed in the next chapter.

5 Test Of Bell's Inequality

> I am sure that Einstein would certainly have had something very clever to say about it.
>
> *Alain Aspect,*
> speculating on how Einstein would have reacted to his experiment.

5.1 Introduction

Bell's landmark paper appeared in 1965. Four years later, Clauser *et al* refined Bell's arguments to suit a practical experiment realisable in the lab. You might wonder why there is so much fuss about a practical experiment. Let me explain. Consider Fig.5.1 which shows a schematic of the EPR experiment (for spin 1/2 particles). Source S emits a pair in the singlet state. The particles then fly off in opposite

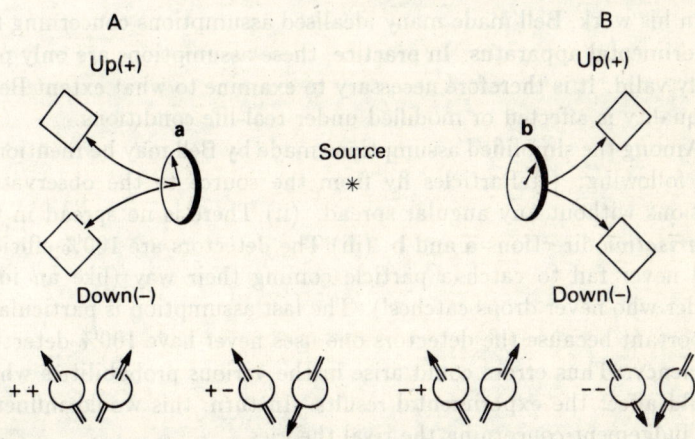

Fig.5.1 Schematic of the EPR experiment for spin 1/2 particles. Various left-right spin combinations are shown. The corresponding probabilities are defined in (5.1) and made use of as in (5.2).

directions towards two observation stations. Measurements on the two particles are made in the directions **a** and **b**. As you already know, each measurement of spin can yield either +1 or −1. We now define four joint probabilities as follows:

$P_{++}(\mathbf{a}, \mathbf{b})$ = probability that spin is +1 along **a** and +1 along **b**.

$P_{--}(\mathbf{a}, \mathbf{b})$ = probability that spin is −1 along **a** and −1 along **b**. (5.1)

$P_{+-}(\mathbf{a}, \mathbf{b})$ = probability that spin is +1 along **a** and −1 along **b**.

$P_{-+}(\mathbf{a}, \mathbf{b})$ = probability that spin is −1 along **a** and +1 along **b**.

The quantity $E(\mathbf{a}, \mathbf{b})$ introduced in the last chapter is related to the probabilities in (5.1) by the relation

$$E(\mathbf{a}, \mathbf{b}) = P_{++} + P_{--} - P_{+-} - P_{-+} \qquad (5.2)$$

Thus, to determine $E(\mathbf{a}, \mathbf{b})$ the four probabilities defined in (5.1) have to be measured and then combined as in (5.2). The other expectation values in Bell's inequality (4.8,9) have to be obtained in a similar manner.

In his work, Bell made many idealised assumptions concerning the experimental apparatus. In practice, these assumptions are only partially valid. It is therefore necessary to examine to what extent Bell's inequality is affected or modified under real-life conditions.

Among the simplified assumptions made by Bell may be mentioned the following: (i) Particles fly from the source to the observation stations without any angular spread. (ii) There is no spread in the *polarisation* directions **a** and **b**. (iii) The detectors are 100% efficient and never fail to catch a particle coming their way (like an ideal fielder who never drops catches!). The last assumption is particularly important because the detectors one uses never have 100% detection efficiency. Thus errors could arise in the various probabilities which would affect the experimental results. In turn, this would influence the judgement concerning the rival theories.

Most of the experiments reported so far have been performed with photons. Figure 5.2 shows the schematic of a photon correlation experiment. A photon pair is produced at S, the photons then moving

Test of Bell's inequality 53

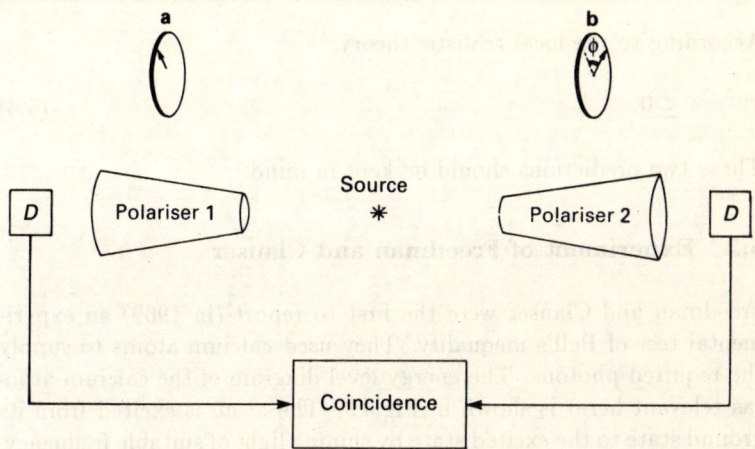

Fig.5.2 Simplified schematic of the photon correlation experiment. The two poralisers define the directions **a** and **b**. The angle between **a** and **b** is denoted by ϕ.

towards the two detector stations on the right and the left. Each station consists of a polariser which can be rotated and also removed if required. Behind the polariser is placed the photon detector. The following quantities are measured:

1. $R(\phi)$ the coincidence rate between the two detectors for a given angle ϕ between **a** and **b**.
2. R_1 the coincidence rate with the polariser 2 removed.
3. R_2 the coincidence rate with the polariser 1 removed.
4. R_0 the coincidence rate with both the polarisers removed.

Next the experimental data is analysed to compute the following:

1. The ratio $\{R(\phi)/R_0\}$.
2. $\delta = |[R(22.5^0)/R_0] - [R(67.5^0)/R_0]| - 1/4$.

These quantities are then compared with theoretical predictions. According to QM,

$$R(\phi)/R_0 = (1/4) + f(\phi).\cos 2\phi \qquad (5.3)$$

where $f(\phi)$ is some function of ϕ. Its precise form is not important for us here.

54 What is reality?

According to the local realistic theory,

$$\delta \leq 0. \tag{5.4}$$

These two predictions should be kept in mind.

5.2 Experiment of Freedman and Clauser

Freedman and Clauser were the first to report (in 1969) an experimental test of Bell's inequality. They used calcium atoms to supply the required photons. The energy level diagram of the calcium atom (as relevant here) is shown in Fig.5.3. The atom is excited from its ground state to the excited state by shining light of suitable frequency.

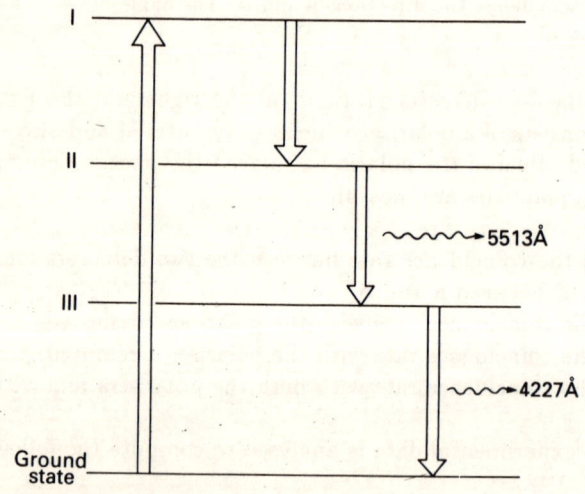

Fig. 5.3 Schematic of a portion of the energy levels of the calcium atom. In the experiment of Freedman and Clauser, the atom was excited to the state I by shining light. From there, the atom cascades down to the ground state via levels II and III, emitting photons. The cascade formed by the photons with wavelengths 5513 Å and 4227 Å were employed. The frequencies are slightly different but that does not matter. These form the pair whose polarisation correlations are tested.

From this excited state the atom returns to the ground state in several steps, in the process emitting photons of wavelength 5513 Å and

4227 Å respectively—see Fig.5.3. The two photons are emitted as a cascade. The experimental arrangement used is essentially as in the schematic shown in Fig 5.2. The ratio $R(\phi)/R_0$ obtained as a function of the angle ϕ is shown in Fig.5.4. The solid line in this figure is the

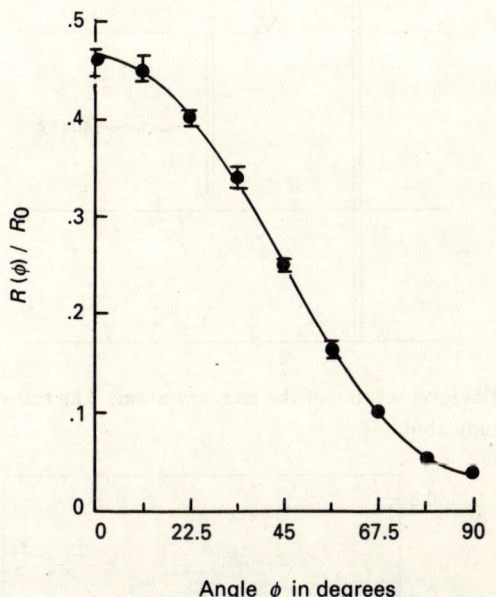

Fig.5.4 $R(\phi)$ versus ϕ as measured by Freedman and Clauser. The solid line is the prediction of QM. The vertical bars indicate experimental errors.

prediction of QM. Clearly QM is doing a good job. Freedman and Clauser also found

$$\delta = 0.050 \pm 0.008$$

compared to the result $\delta \leq 0$ predicted by a hidden-variables type of theory. The experiment of Freedman and Clauser definitely favours QM.

5.3 Experiment of Clauser

In 1976, Clauser reported a photon correlation experiment performed with a cascade from the mercury atom. The cascade used by him is shown in Fig.5.5. The rest of the experiment was quite similar to that performed by him earlier with Freedman. Clauser performed this

56   *What is reality?*

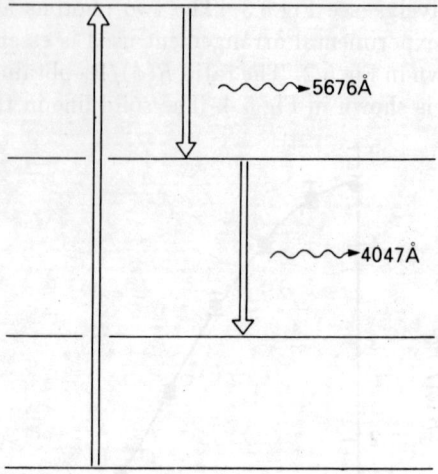

Fig.5.5 Partial level scheme of the mercury atom. The transitions used by Clauser are indicated.

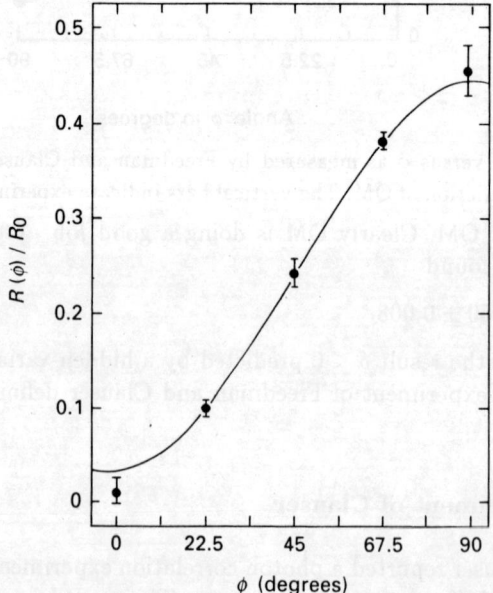

Fig.5.6 $R(\phi)$ plot obtained by Clauser. The solid line is the prediction of QM.

experiment because in 1973, Holt and Pipkin did an experiment using radiation from mercury and found that their results did not support QM. In other words they found that Bell's inequality is obeyed. Naturally this was disturbing news which is why Clauser decided to repeat the Holt–Pipkin experiment. Clauser found that the ratio $R(\phi)/R_0$ varied with ϕ as shown in Fig.5.6; there was no disagreement with QM. He also obtained for δ the value 0.0385 ± 0.0093, clearly violating the inequality (5.4). On the other hand, QM predicts $\delta = 0.0348$. So one more vote for QM. But this raises the question why Holt and Pipkin got a different result. Did they go wrong somewhere? The answer is not yet known.

5.4 Experiment of Fry and Thompson

Fry and Thompson also wished to repeat Holt and Pipkin's experiment and like the latter, they too used radiation from the mercury atom. Their result for the ratio $R(\phi)/R_0$ is shown in Fig.5.7. Once again the agreement with QM is very good. Any lingering doubts are removed by the value obtained for δ by Fry and Thompson. Their result is $\delta = 0.046 \pm 0.014$, in excellent agreement with the prediction

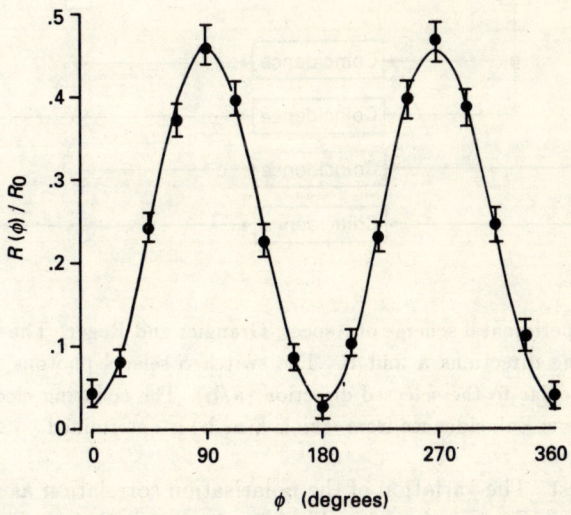

Fig.5.7 $R(\phi)$ plot observed by Fry and Thompson. The solid line is the prediction of QM.

from QM which is $\delta_{QM} = 0.044 \pm 0.007$, and in clear violation of the hidden-variables type of theory.

5.5 Experiment of Aspect, Grangier and Roger

The reason for this experiment can be understood by going back to (5.2). To obtain $E(\mathbf{a}, \mathbf{b})$, one needs to measure the four probabilities defined in (5.1). In the experiments discussed thus far, these probabilities could not be directly measured on account of practical limitations. Instead one went about the whole business in a somewhat roundabout manner by removing one polariser, then the other and so forth. Aspect *et al* employed the arrangement shown in Fig.5.8. Here nothing is removed and all the required combinations, i.e., $++$, $--$, $+-$ and $-+$ are properly measured. Bell's inequality is therefore given

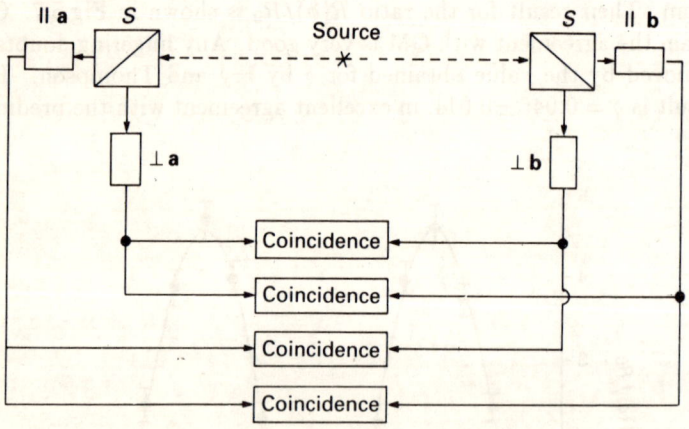

Fig.5.8 Experimental scheme of Aspect, Grangier and Roger. The polarisers select the directions **a** and **b**. The switch S selects photons parallel or perpendicular to the selected direction (a/b). The counting electronics records various coincidences from which $E(\mathbf{a}, \mathbf{b})$ is determined.

a proper test. The variation of the polarisation correlation as a function of the angle ϕ is shown in Fig.5.9. As usual, QM comes out a winner. Aspect *et al* also determined the value of the quantity

$$S = E(\mathbf{a}, \mathbf{b}) - E(\mathbf{a}, \mathbf{b}') + \mathbf{E}(\mathbf{a}', \mathbf{b}) + \mathbf{E}(\mathbf{a}', \mathbf{b}'). \tag{5.5}$$

The measured value is $S = 2.70 \pm 0.05$ whereas QM predicts $S = 2.697 \pm 0.015$. Could one ask for anything better?

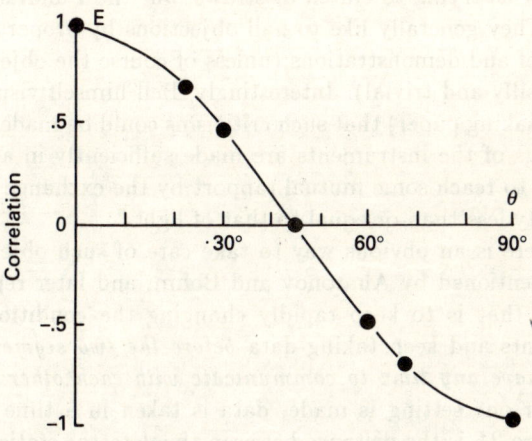

Fig.5.9 $R(\phi)$ data of Aspect *et al* and the quantum mechanical prediction.

5.6 Test of nonseparability

The story is not over yet. One objection can be raised against all the experiments discussed so far, including "the new improved one" reported by Aspect and coworkers. As I told you earlier, when Bell derived the inequality which should be obeyed by a hidden-variables theory, he essentially made three assumptions. Of these, I would like to call attention now to the assumption of locality. This is also called separability or Einstein separability for emphasis. Reduced to brass tacks, what it means in the present context is that any setting made to one of the two measuring apparatuses should not be able to instantly influence the other apparatus. If at all there is to be any influence, it can at best be by the propagation of a signal that travels with the speed of light. In the experiments discussed thus far, how can one be certain that the two instruments did not enter into some sort of a conspiracy in the past? It might well be that QM is getting support on account of such prior adjustments.

You see how suspicious one gets in this whole business! Let us pin down the objector and ask him to be more specific about the conspiracy theory. The objector replies: "When you set one instrument at the beginning of the experiment, it might send signals to the other instrument and affect the events observed there or it could modify the hidden parameters at the source. As a result of all this, you are getting results favourable to QM." You and I might be inclined to dismiss the objector as trying to clutch at straws, but the Pundits are more cautious. They generally like to nail objections by proper argumentation, proof and demonstrations (unless of course the objections are downright silly and trivial). Interestingly, Bell himself visualised (in his epoch making paper) that such criticisms could be made, i.e., that "the settings of the instruments are made sufficiently in advance to allow them to reach some mutual rapport by the exchange of signals with velocity less than or equal to that of light."

Well, there is an obvious way to take care of such objections. It was first mentioned by Aharonov and Bohm, and later repeated by Bell. And that is to keep rapidly changing the conditions of the measurements and keep taking data *before the two segments of the apparatus have any time to communicate with each other.* In other words, after one setting is made, data is taken in a time less than $(2L/c)$ where $2L$ is the distance between the detector stations on the two sides of the source.

In 1976, Aspect proposed the scheme for such an experiment, and in 1982 he actually carried out the experiment along with Dalibard and Roger. By the way, these experiments were performed in Paris. The arrangement employed by them is shown in Fig.5.10 and employs

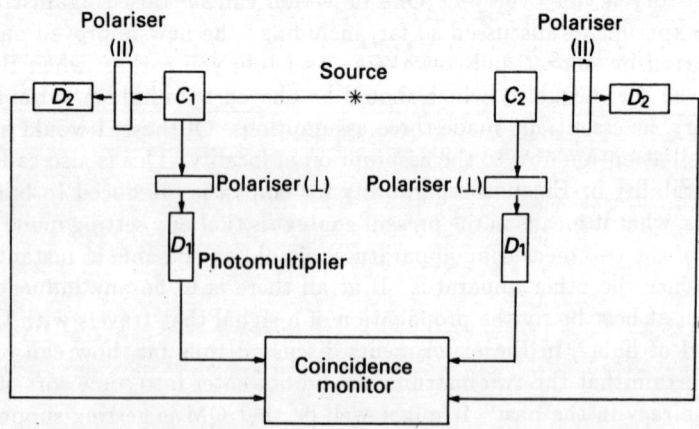

Fig.5.10 Arrangement used by Aspect *et al* in their rapid switching experiments. Commutators C_1 and C_2 rapidly switch between polarisers parallel and perpendicular. They work independently. Various coincidences are measured as in Fig.5.8 and Bell's inequality tested.

two switches C_1 and C_2, which rapidly switch the photons from one polariser–detector combination to another. Aspect *et al* refer to these switches as commutators. Don't confuse these with quantum mechanical commutators like $[\hat{p}, \hat{q}]$! They are more like the commutators used in electrical engineering to switch circuits. The details of the commutator are shown in Fig.5.11. Basically, one has a cell containing water. An acoustic wave is propagated through the water creating a

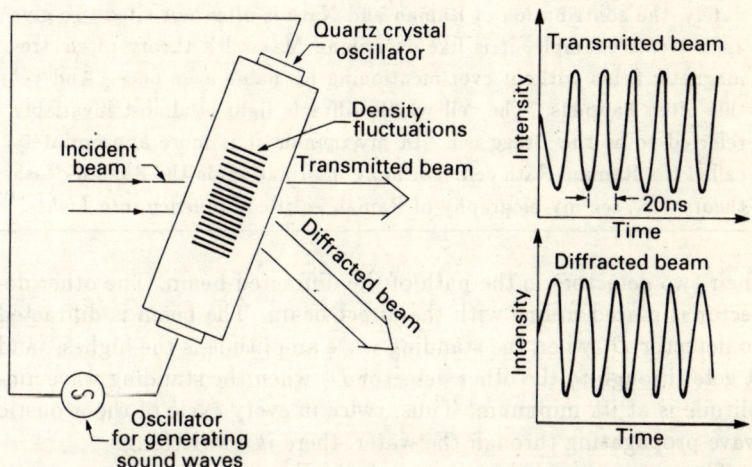

Fig.5.11 Optical commutator. It consists of a Raman–Nath cell (see Box 5.1) in which acoustic standing waves are set up. The density variations produce a grating which diffracts the light beam. However, as the amplitude of the wave varies twice every acoustic cycle, diffraction also occurs twice every cycle when the amplitude is maximum. At the minima, the optical wave passes direct without deflection.

standing wave which produces regions of compression and rarefaction in the water. The refractive indices are different in the two regions. The net result of all this is that when the standing wave has its maximum amplitude, the water cell behaves like a diffraction grating (see also Box 5.1). Light going through the cell is then diffracted. Aspect and his coworkers take advantage of this and put one of

> **Box 5.1** Light is diffracted whenever there is a periodic disturbance. In the late twenties, it was demonstrated in experiments in Europe and in America that when standing waves are set up in a liquid, the liquid acts as a grating and diffracts light. There is nothing surprising about this.

> What could not be understood, however, was the peculiar manner in which the intensity of the diffracted light varied when the amplitude of the impressed sound wave was altered. In 1935, Raman and Nagendra Nath gave a theory for this anomalous behaviour of the diffracted intensity. They wrote in all four papers on the subject and the theory is known as the Raman–Nath theory. It has withstood the test of time. Today it is used all over, including technological applications. Unfortunately, the contribution of Raman and Nath is often not cited. To give an extreme example, it is like describing Maxwell's theory of electromagnetic fields without ever mentioning his name even once. And yet this often happens. The cell which diffracts light is almost invariably referred to as the Bragg cell. In my opinion it is more appropriately called the Raman–Nath cell. For more information on the Raman–Nath theory etc., see my biography of Raman entitled, *Journey into Light*.

their two detectors in the path of the diffracted beam. The other detector is placed in line with the direct beam. The beam is diffracted to detector D_1 when the standing wave amplitude is the highest, and it gets through to the other detector D_2 when the standing wave amplitude is at its minimum. Thus, twice in every cycle of the acoustic wave propagating through the water, there is a switching.

The two commutators are switched independently by two crystals with different frequencies which have no phase relationship to each other—not as good as random switching but it is nearly pseudorandom, which is the next best.

In the experimental arrangement used by Aspect and his colleagues, each detector station was 6 m from the source. The switching between two channels occurred roughly every 10 nanoseconds (1 ns = 10^{-9}s). If a signal were to propagate from one end of the apparatus to the other, it would require 40 ns which means that in this experiment, the detectors have no time whatsoever to conspire. The time sequence of switching and detection are schematically illustrated in Fig.5.12. Now for the result. The usual correlation plot is given in Fig.5.13. Aspect *et al* also reported a quantity 's' for which experiment gives the result 's'= 0.101 ± 0.020, which is to be compared with the quantum mechanical prediction 's' = 0.112 and the classical result 's' ≤ 0 *derived using the idea of Einstein separability.*

The result obtained in this experiment spells disaster for the hidden-variables type of theories. Clearly, their predictions do not agree with experiment which means that the underlying assumptions must be questioned, that is, objective reality and local separability must be questioned—particularly the latter. Einstein would have been most unhappy but there it is!

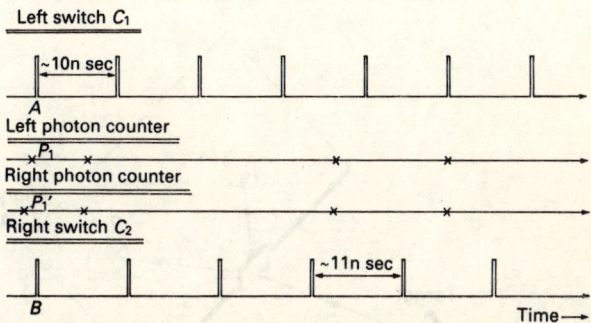

Fig.5.12 C_1 and C_2 show the switchings sequence of the two commutators. Their frequencies are not identical. Consider the photon pair P_1, P_1'. These are detected on the left and right after the switchings A and B respectively. Notice that after A is switched, P_1' is detected before a signal can be transmitted from the left to the right. Similarly, P_1 is detected before the switching of B can be communicated.

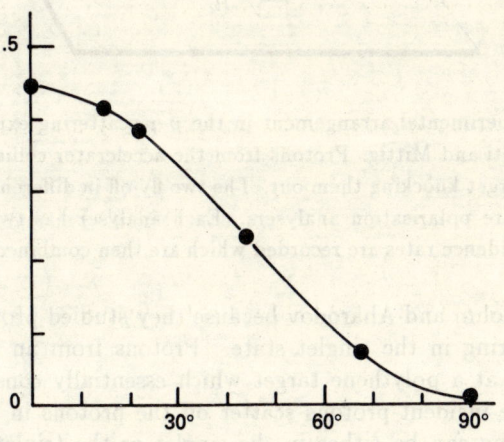

Fig.5.13 The correlation plot as obtained by Aspect, Dalibard and Roger.

5.7 Experiment of Lamehi–Rachti and Mittig

I have been rather partial to photon correlation experiments. This however does not mean that other particles have not been used. In fact they have been as the experiment to be now described will show. The experimental arrangement used by Lamehi-Rachti and Mittig is shown in Fig.5.14. Their scheme comes rather close to the original

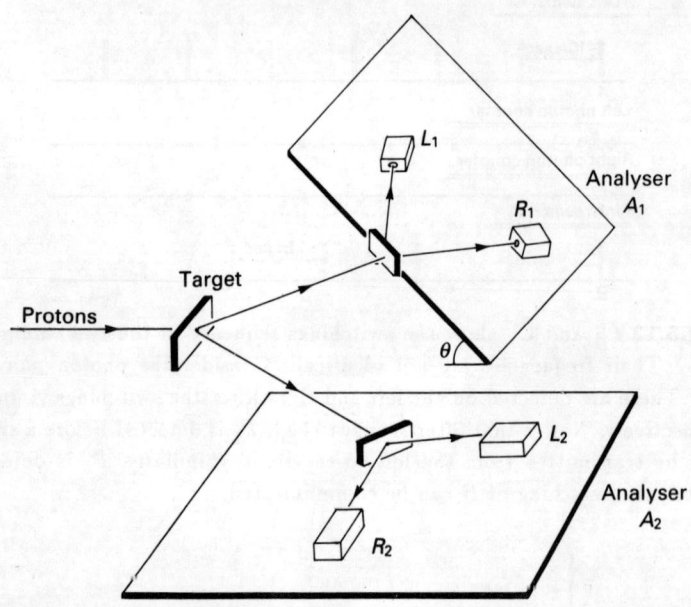

Fig.5.14 Experimental arrangement in the p-p scattering experiments of Lamehi-Rachti and Mittig. Protons from the accelerator collide with protons in the target knocking them out. The two fly off in different directions. A_1 and A_2 are polarisation analysers. Each analyser has two detectors. Various coincidence rates are recorded which are then combined as in (5.6).

concept of Bohm and Aharonov because they studied proton-proton (p-p) scattering in the singlet state. Protons from an accelerator are directed at a polythene target which essentially consists of hydrogen. The incident protons scatter off the protons in the target. The scattering can be either in the singlet or the triplet state but the conditions of the experiment were chosen to favour the singlet state. After scattering, the two protons fly off in different directions as shown. The spin states of the two protons are determined using the polarisation analysers A_1 and A_2. Each analyser consists of a carbon target from which the proton is scattered either to a "right" detector (R) or a "left" detector (L).

The experiment consists in fixing the relative angle θ (see Fig.5.14) at a particular value and measuring the various coincidence rates, i.e., $L_1L_2, L_1R_2, R_1L_2, R_1R_2$. Call these as $N(LL), N(LR), N(RL)$ and $N(RR)$ respectively. The measured spin correlation $P(\theta)$ is defined as:

$$P(\theta) = \frac{N(LL) + N(RR) - N(RL) - N(LR)}{N(LL) + N(RR) + N(RL) + N(LR)} \quad (5.6)$$

The results obtained are shown in Fig.5.15.

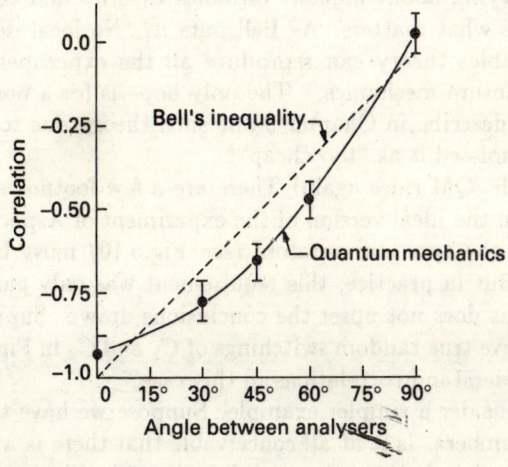

Fig.5.15 $P(\theta)$ as measured by Lamehi-Rachti and Mittig.

The points correspond to experimental observations and the vertical bars denote the experimental errors as usual. The two continuous lines have been drawn on the basis of QM predictions and Bell's inequality. As expected, there is clear evidence of violation of Bell's inequality in a certain region.

5.8 The score

As you have just seen, practically all experiments agree with QM. So where does it leave the hidden-variables type theory? In a tight corner I'm afraid! In fact, an article which appeared in the journal *Science* soon after Aspect's experiment, started with the words: "Local hidden-variables theory is dead."

For a minute, forget QM, its predictions, etc., and let us concentrate on the disagreement between the hidden-variables theories and experiment. Normally, when there is disagreement between the predictions of theory and experimental results, one tries to modify the theory suitably. In particular, one tries to modify the starting assumptions. In this case, this question has been looming large even before Aspect's latest experiments; the experiment of Aspect, Dalibard and Roger has complicated matters even further. Does one construct a hidden-variables theory by giving up objective reality (we can't do that!) or by giving up Einstein separability (that too must remain) or inductive logic itself? No one knows; the simplest thing is to stop worrying about hidden- variables theories and concede that QM alone is what matters. As Bell puts it, "No local deterministic hidden-variables theory can reproduce all the experimental predictions of quantum mechanics." The only hope is for a non-local theory. I shall describe in Chapter 9 one such theory due to Bohm but Einstein dismissed it as "too cheap"!

Meanwhile, QM rides again! There are a few footnotes that must be added. In the ideal version of the experiment of Aspect, Dalibard and Roger, the two commutators (see Fig.5.10) must be switched randomly. But in practice, this requirement was only partially met. However, this does not upset the conclusions drawn. Suppose we are able to achieve true random switchings of C_1 and C_2 in Fig.5.10. How does one understand correlations in this case?

Let us consider a simpler example. Suppose we have two random strings of numbers. Is it at all conceivable that there is a correlation between them? Indeed this is possible as is evident by comparing the strings 31415926535897 ... and 20304815424786 ... Each string is a random medley of numbers but if we subtract the second string from the first integer by integer, we obtain the sequence 111111111111 ... which is definitely not random. The moral is that apparently random patterns can have non-random correlations depending upon their origin. This has a message for us. Let us hear John Gribbin on this subject. Commenting on Aspect's experiment (the last one), he says:

> It tells us that particles that were once together in an interaction remain in some sense parts of a single system which responds together to further interactions. Virtually, everything we see and touch and feel is made up of collections of particles that have been involved in interactions with other particles right back through time to the Big Bang...

D'Espagnat makes a similar observation:

> Most particles or aggregates of particles that are ordinarily regarded as separate objects have interacted at some time in the past with other objects. The violation of separability seems to imply that in some sense all these objects constitute an indivisible whole.

So, what is the bottom line? If we consider a pair of electrons born say in the singlet state and the electrons fly off in opposite directions, they are "bound together by the non-local" wave function

$$\Phi(1,2) = (1/\sqrt{2})\{\Psi_{+-} - \Psi_{-+}\}$$

which I introduced you to in (4.5); this is so till we decide to observe either electron 1 or 2. The moment we make an observation on one of them, Φ collapses to Ψ_{+-} or Ψ_{-+}. In both cases, the two electrons somehow "know" what to do. If spin of 1 is up $(+)$, the spin of the other is automatically down $(-)$ or vice versa, just like in Bertlmann's socks. The beauty is that the two electrons instantly know what to do, even if they are at the two ends of the Universe. This is, if you like, cooperation without communication! So relativity is not violated—this is the way it is all explained.

All this is quantum mechanical wisdom, and whether we like it or not, experiments specially designed to test these very features show that QM is OK. Although one might use words like correlations, entanglement, etc., I am sure you are quite uneasy about this business of electrons somehow instantly knowing what to do, even though they are far far apart. You might even complain that there is some hocus-pocus going on. I wouldn't blame you. There is an explanation of sorts which you will find in Box 5.2.

Box 5.2 Here is a slightly more quantitative way of understanding what Gribbin and d'Espagnat say. This is based on remarks by Bell. I have to use the space-time description with which you must be familiar, at least from Part II (if not from the companion volume *At the Speed of Light*). E_1 and E_2 are simultaneous events as shown in figure (a). Obviously signals relating E_1 and E_2 cannot be transmitted from x_1 to x_2 or vice versa, unless it is done with infinite speed which, however, we do not permit. But now consider the backward light cones associated with these events. They overlap. Thus in the distant past there could have been a tie up of some sort which leads to correlations at time t. "Ancient conspiracy", if you like, not of the parts of the apparatus but of the primary causes responsible for the events.

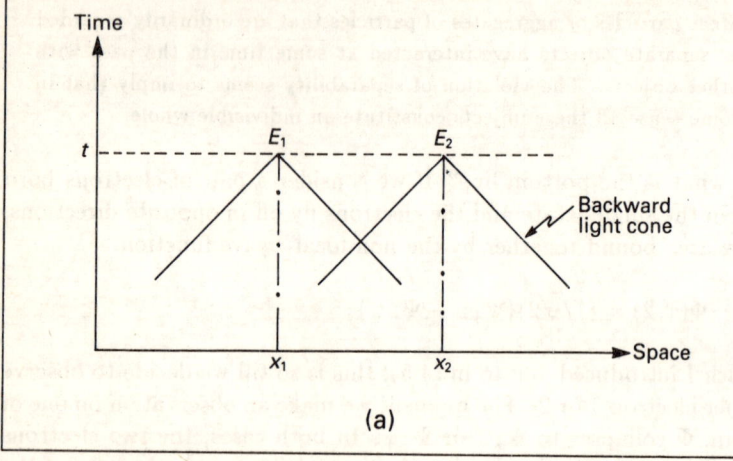

(a)

However, Bell himself says:

> In these EPR experiments there is the suggestion that behind the scenes something is going faster than light ... It is as if there is some kind of a conspiracy, that something is going on behind the scenes which is not allowed to appear on the scenes. And I agree that's extremely uncomfortable.

Well, that is all the comfort I can offer you!

6   *On Measurements*

> I am convinced that the word 'measurement' has now been so abused that the field would be significantly advanced by banning its use altogether, in favour for example of the word 'experiment'.
>
> J.S. Bell

As I have described in the last chapter, experiments have given a clear vote in favour of QM. However, the underlying philosophical picture is by no means clear. There are two views about reality. One is the age old classical picture which Einstein strongly supported and which also is in accord with our natural instincts. This picture definitely operates in the macro world and may therefore also be referred to as macro-reality. As opposed to this there is quantum reality, forced on us by the successes of QM.

QM is normally applied to the micro-world and quantum reality is therefore usually taken to be synonymous with micro-reality. Thus the two pictures were applied in different realms and did not produce serious qualms especially because we lack direct experience of the micro-world. Trouble arose when one applied QM to the macro-world, as we have seen in the case of the cat paradox.

According to QM, the state of the cat in the box is $|\text{Alive}\rangle + |\text{Dead}\rangle$, which of course makes no sense at all to us. But when one performs a measurement (i.e., opens the box), one finds either a live cat or a dead one, exactly as per our (classical) expectations. In some mysterious manner, measurement transforms a QM state into one which makes classical sense. There is no clear cut prescription regarding how this quantum picture changes into classical reality; all we can do is to wave hands and mutter something about wave functions collapsing.

Measurement plays its central role in QM. In fact, quantum mechanics is fundamentally about measurements. It necessarily divides the world into two parts, a part which is observed and a part that does the observing; we usually refer to the latter as the apparatus. Further, we suppose that while the first part exhibits quantum behaviour, the part that does the observing is classical — this inspite of the fact that the apparatus is itself always built up from quantum systems (atoms), though maybe a very large number of them.

One problem that keeps coming up again and again is: Where do we draw the line between the system and the apparatus? You might remember (see Chapter 2) that Bohr shot down the EPR argument essentially by taking the slit as a part of the observer or the apparatus when one measures the momentum and as a part of the observed system when one measures the coordinate. So you see that this boundary has to be drawn with care. In fact, we do not even know if the line is sharp or blurred.

I am now going to complicate matters further by asking whether we include a person when we talk of the apparatus, measurements and so forth. After all, in QM we do often refer to an observer, and normally an observer implies a living person. Before we get into the mind and consciousness business, a few words about the "orthodox" view. The standard viewpoint is that by "observer" is meant an inanimate apparatus which amplifies microscopic events to macroscopic consequences. Paraphrasing Bohr, John Wheeler says:

> No elementary quantum phenomenon is a phenomenon until it's brought to a close by an irreversible act of amplification by a detection ...

And Bell adds:

> Of course this apparatus, in laboratory experiments, is chosen and adjusted by the experimenters. In this sense the outcomes of experiments are indeed dependent on the mental processes of the experimenters! But once the apparatus is in place, and functioning untouched, it is a matter of complete indifference ... according to quantum mechanics ... whether the experimenters stay around to watch, or delegate such 'observing' to computers. ... Our apparatus visits the microscopic world for us, and we see what happens to it as a result.

At this stage, I must introduce you to what is called in technical literature as the problem of *infinite regression*. The problem was apparently first brought to the surface by the famous mathematician von Neumann (whose name is also well known in computer science) when in the thirties, he did a detailed study of the mathematical foundations of quantum mechanics.

In qualitative terms, the problem is roughly the following: In QM_1 observation is king. To observe, we need an apparatus. Only when an observation is made does one of the various possible potentialities become a physical reality. Fine so far. Now the apparatus is supposed to be classical. However, it is itself made up of atoms and as we know, atoms are quantum systems. So how does the apparatus become a reality? Only when observed with another apparatus. When does that apparatus become a reality? Only when observed with yet another apparatus. And so on ...

I hope you are getting the point. An atomic system can be in a superposed or "schizophrenic" state as in (3.2). To make it decide, we use an apparatus (like the cat). But this apparatus now gets into a superposed state. To break that we use another apparatus, and another apparatus and so on. Where will it all end?

There is one answer to this which I have already indicated in Chapter 3 which is that the process ends when information is generated in the macroscopic world accompanied by an increase in the entropy of the Universe (whatever all that means!). Another way of breaking this chain was suggested by Wigner, which is what I would like to discuss now.

There is one other ticklish issue that must be referred to before I bring the mind into the picture. Suppose the Universe itself is the system to be observed. What do we do now about the apparatus? Do we put it "outside" the Universe? What does that mean? Or do we introduce a live observer into the inanimate Universe for doing the "observing"? I am not sure if all these questions have been satisfactorily answered.

We now introduce living objects into the picture and allow them to come into contact with the quantum system. All sorts of questions come up immediately: Can a single cell (the simplest example of a living object) be given the status of an observer or must the organism be something more developed than a cell? Could it be a fly or a cat or must it be a human who has what we call a *mind*? As Bell puts it:

> What exactly qualifies some subsystems to play this role? Was the world wave function waiting to jump for thousands of millions of years until a single-celled living creature appeared? Or did it have to wait a little longer for some more highly qualified measurer — with a Ph.D.?

Enter now the mind. Things immediately get even more difficult because it raises the question about the role of *consciousness* in QM.

72 What is reality?

So you see how complicated the whole thing gets. Perhaps you will now appreciate why I remarked in Part I that QM raises deep and philosophical questions. Of course it is a different matter, as someone observed, that the average quantum mechanic is as little concerned about such issues as is the garage mechanic! But not everyone can afford to take such a negligent view.

In ancient times, many philosophers believed that *the mind is primary*. This led to a type of philosophy called *solipsism* according to which the self is the only thing that can be known to exist. Starting from the sixteenth century onwards solipsism went into a retreat, thanks to the brilliant success of mechanics and various other branches of physics. Chemistry of course made its own contributions to this eclipse. As a result, there evolved a *mechanistic* view of the Universe. With the advent of quantum mechanics, however, things began to change. Commenting on this change of perspective, Heisenberg once declared:

> The laws of nature which we formulate mathematically in quantum theory deal no longer with the particles themselves but with *our* knowledge of the elementary particles ... The conception of objective reality ... evaporated into the mathematics that no longer represents the behaviour of elementary particles but rather our knowledge of this behaviour.

Eugene Wigner feels that the word "our" used by Heisenberg refers to the observer, and according to him the "live" observer has to be taken seriously since in QM we deal with an epistemological point of view (maybe you would like to refresh your memory about the definition of this word). Wigner adds:

> Through the creation of quantum mechanics, the concept of consciousness came to the fore again: it was not possible to formulate the laws of quantum mechanics in a fully consistent way without reference to the consciousness.

I realise you must be feeling absolutely dizzy with this big jump from wave functions to consciousness but perhaps you appreciate at least now that QM *has* opened new vistas in philosophy. Let us listen to Wigner again. He says:

> Given any object, all possible knowledge concerning that object can be given as its wave function ... The information given by the wave function is communicable. If someone else somehow determines the wave function of a system, he can tell me about it ...

According to Wigner, in this sense the wave function "exists". He adds, "all knowledge of wave functions is based, in the last analysis, on the 'impressions' we receive." Remember the earlier discussions and remarks about wave function collapse ? Wigner now links it all to consciousness! In other words, the wave function collapses and the regression is ended when information enters into consciousness. Wigner says:

> The impression which one gains at an interaction [with a quantum system], called also *the result of an observation* , modifies the wave function of the system. The modified wave function is, furthermore, in general unpredictable before the impression gained at the interaction has entered our consciousness: it is this entering of an impression into our consciousness which alters the wave function ...

So you see how consciousness has been brought into the picture. In brief, according to Wigner if one speaks in terms of the wave function, "its changes are coupled with the entering of impressions into our consciousness". A philosopher of old would perhaps wryly comment that we are returning to solipsism, at least in a limited fashion.

One might wonder what if someone else other than us makes the observation? Bell gives an interesting twist to this question. Suppose a scientist performs an experiment and reports the result in a scientific journal say *The Physical Review*. Let us say another scientist (whom we shall call the reader) reads the report in *The Physical Review*. As far as the reader is concerned, an impression is formed in his mind only at the time he *actually* reads the article. Does it mean that the experimenter, his equipment, the Editor of *The Physical Review*, the referee, the printer, the mailman, etc., all form a part of the overall "apparatus" and that the "wave function collapses" only at the time the reader reads the paper? This whole business is getting weird, isn't it?!

Roger Penrose has his own version of such a complication. He modifies Schroedinger's cat experiment by putting the cat in a room which can also accomodate a person. Of course this person is let into the room only with suitable protective clothing, gas mask, etc. So we are outside the room while this other person is inside. There is now a peculiar situation. Supposing that the source actually emits a particle and that as a result the cat dies. The observer inside would know about it and for him the wave function would have collapsed. We, on the other hand, are outside. The door is still closed and as far we are concerned, the cat is in a linear superposition of being dead and alive! How do we untangle this messy situation? Wigner who has thought about such matters offers the following clarifications.

He starts with three entities: The quantum system, a "friend" and ourselves. The quantum system is an atom which is capable of being in one of two states: ψ_1 in which it has not emitted light and ψ_2 in which it has emitted a flash. We next consider the atom and the friend as a composite system. If the atom is initially in the state ψ_1, then when the friend interacts with the atom, the composite system would be in the state $\psi_1 \times \chi_1$. The state χ_1 of the friend implies that if he is asked the question, "Did you see a flash (from the atom)?" he would reply "No". If the atom was in the state ψ_2 to start with, the composite system would be described by $\psi_2 \times \chi_2$. The state χ_2 of the friend now is such that in reply to our question he would say, "Yes".

So far there is no problem. Suppose the atom is in the superposed state $(\alpha\psi_1 + \beta\psi_2)$ to start with. When the friend interacts with the atom, we expect, on the basis of the linear equations of QM, the state of the composite system to be described by

$$\alpha(\psi_1 \times \chi_1) + \beta(\psi_2 \times \chi_2) \tag{6.1}$$

If now we ask the friend whether he saw a flash or not, there is a probability $|\alpha|^2$ he would say "No" and a probability $|\beta|^2$ he would say "Yes". All this is quite satisfactory.

Trouble starts if after the experiment we decide to have a cup of tea with our friend and discuss the whole business. We ask, "I say old chap, what did you feel about the flash *before* I asked you the question?" He will reply, "I told you already that I did not (did) see a flash" as the case may be. This is interesting because we learn that our friend knew the state of the atom even *before* we had questioned him. Which means that the state of the composite system was not as in (6.1) but was $\psi_1 \times \chi_1$ or $(\psi_2 \times \chi_2)$ as the case may be. On the other hand, if instead of a "friend" we had an inanimate object doing the observing, then the wave function of the composite system would be as given by (6.1). In other words, while an inanimate object can be in a "state of suspended animation", a live person cannot be. This leads Wigner to conclude "that the being with a consciousness must have a different role in quantum mechanics than the inanimate measuring device."

Wigner's views are not necessarily accepted by everyone but I brought them to your notice just to highlight the serious speculations engaged in by some of the big shots.

In an interview over the BBC, John Wheeler has commented on

Wigner's ideas. Wheeler feels (along with the Norwegian philosopher Follesdal) that observation acquires meaning or becomes knowledge only when it is communicated. Communication, then, is the essential idea. Here is an excerpt from the interview:

> *If we turn back to the second stage of the measurement process — the establishment of knowledge — I can't help feeling that this is rather like Wigner's interpretation of quantum theory, that the translation from the quantum phenomenon to knowledge or meaning depends on the existence of conscious observers. Is that right?*

Wigner speaks of the elementary quantum phenomenon as not really having happened unless it enters the consciousness of an observer. I would rather say that the phenomenon may have just happened but may not have been put to use. And it is not enough for just one observer to put it to use — you need a community.

Nevertheless, you still regard the existence of conscious observers as crucial to that second stage.

That's right. Although this word conscious is a little tricky here because one can think of animals as having brains that are primitive ... So I would not like to put the stress on consciousness even though that is a significant element in this story.

According to Follesdal's statement, meaning is the joint product of all the evidence that is available to those who communicate. So it's the idea of communication that's important. As animals have to communicate, the establishment of meaning doesn't require the use of English!

So is there a distinction here which hinges crucially on the difference between living creatures and inanimate objects?

Yes, this is a most difficult question of where we draw the line.

I don't think the last word on this subject has been said. Surely there will be more speculation but it is also clear that this is a game for "adults" and not "kids"! Let me end this chapter with the following remark by Bell:

> It remains a logical possibility that it is the act of consciousness which is ultimately responsible for the reduction of the wave packet.

7 Complementarity Questioned

> They [the philosophers] did not see that it [the complementarity description of quantum theory] was an objective description and that it was the only possible description
>
> Niels Bohr
in an interview recorded just a few hours before his death.

> In recent years, I have been led to regard the concept of complementarity with increasing suspicion.
>
> Louis de Broglie (in 1964)

7.1 Complementarity recalled

In the famous epic *Mahabharatha*, the venerable Bhishmacharya plays a key role as the father figure. In the history of quantum mechanics, Niels Bohr plays a somewhat similar role. By the early thirties he had laid the foundations for the philosophical interpretation of quantum mechanics. Known popularly as the *Copenhagen interpretation*, this is the "Bible" by which generations of physicists have sworn, even if they did not always understand what it actually meant! At the heart of this interpretation lies the famous *Complementarity principle*. Applied to the wave-particle duality, what this principle effectively says is the following:

> What you see depends on the kind of experimental arrangement used. Any given arrangement or scheme can reveal only one aspect of duality; the complementary aspect is *excluded*. For example, if you want to see the particle-aspect of an electron, you must use one arrangement whereas if you want to see the wave-aspect of the electron then you must use some other arragement. No one arrangement can reveal both the particle and the wave-aspect at the same time.

Actually, Bohr talked about three types of complementarity: (i) Between the wave and the particle pictures, (ii) between position and momentum and (iii) between "space-time coordination" and causal description — the last mentioned has been referred to earlier in Chapter 1. In all cases, one can study only one aspect with a (properly) given experimental arrangement.

This one-aspect-at-a-time business is often called *mutual exclusivity* (ME), and the example invariably cited as an illustration is the famous double-slit experiment with which I started this trilogy (see Chapter 1, Part I). This is ME in practice and so far, practically everyone has been faithfully echoing what Bohr said by citing this example as evidence for the complementarity principle. Over the years, this principle has become a matter of faith and no one has dared to question it.

Now in physics, a principle enjoys a special status. Consider, for example, the Pauli exclusion principle. It has been around for over sixty years and has been invoked on innumerable occasions but there has not been a *single* instance when one had to plead for an exception to the principle; that is the sort of firmness one expects from a principle of physics — no ifs, buts and exceptions. I do not know if the complementarity principle has been invoked so often as the Pauli principle but certainly no one has questioned it, at least until recently. In this chapter I would like to tell you something about the doubts that are now being raised. What is pleasing is that some of these questions have been asked here in India.

7.2 About tunnelling

The counter-example to ME that has been proposed centres around the quantum phenomenon of tunnelling. I must therefore say something more about tunnelling than I did in Part I (see Chapter 3, section 16). Let us start with Fig. 7.1(a) which shows a *classical* particle in a one-dimensional box. By this I mean that the particle has freedom to move only in the $\pm x$ directions. However, since the box has infinite walls at $+a$ and at $-a$, the particle is confined to the region $-a \leq x \leq +a$. In Fig.7.1(b) we again have a classical particle but this time it has to face only one wall, i.e. at $x = +a$. So the particle is restricted on the positive x-axis side only and not on the negative side. There is a further relaxation in Fig.7.1(c) where the wall (at $x = +a$) now has a finite height H. If energy is plotted along the vertical axis, then it follows that for particle energy $E > H$, the particle is completely free, i.e., its range is $-\infty \leq x \leq +\infty$. For $E' < H$, however, the range is $-\infty \leq x \leq +a$.

All this preamble is for leading up to Fig.7.1(d) which shows a barrier of finite height H and a finite width Λ. In classical physics, the finite width makes no difference; in other words, the situation is the same as in the case of Fig.7.1(c). But in the case of QM there is a dramatic difference in that the particle can tunnel through the barrier even if its energy is less than H — see Fig.7.1(e) (recall also Figs.3.11 and 3.12 in Part I).

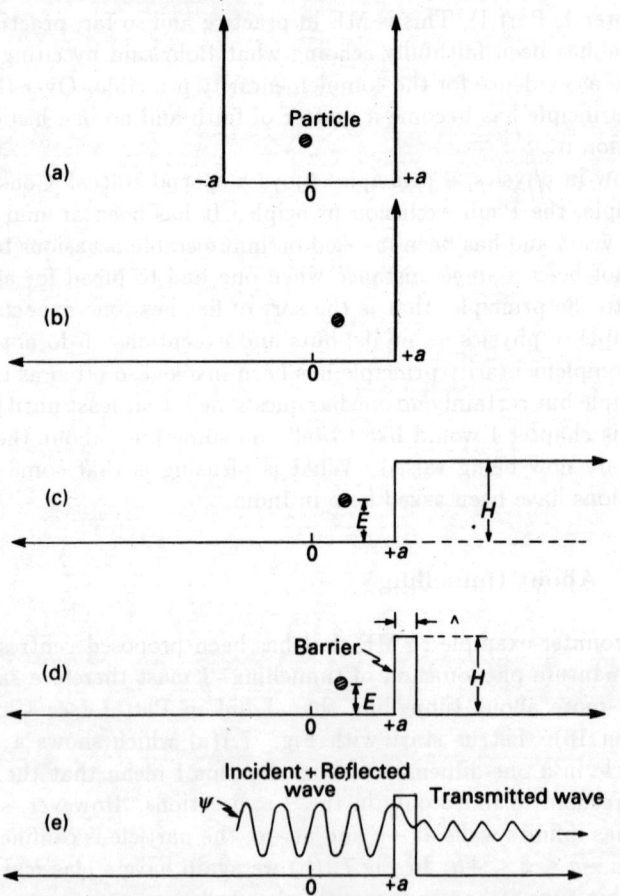

Fig.7.1 (a) Classical particle in a one-dimensional box, the particle is completely confined. (b) The particle now faces only one wall; correspondingly, it is restricted only on the $+x$ side. (c) The wall at $+a$ has a finite height. Thus confinement occurs only for $E < H$. (d) In this case, the barrier has a finite width. However, as far as classical physics is concerned there is no difference compared to (c). (e) shows quantum tunnelling. This is due to the wave nature of the particle.

What is the probability of tunnelling? That depends in general on (i) the particle energy E, (ii) the barrier height H, and (iii) the barrier width Λ. If Λ is small enough, there is a good chance for the particle to leak through; the smaller the value of Λ, the greater is the probability of tunnelling.

Classical physics does have examples of such "leakage", though for waves than for particles. The simplest example one can think of is the passage of light through smoky glass. As shown in Fig.7.2, if the glass slab is thin enough there is some chance for the incident light to filter through. Taking this as a cue, *particle tunnelling in QM is attributed to the wave-aspect of the particle.*

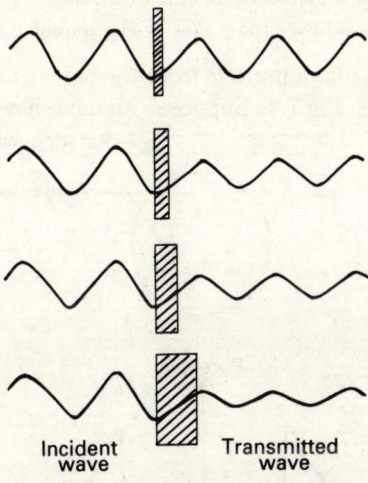

Fig.7.2 Transmission of light through absorbing glass. As can be seen, light leaks through when the slab is thin enough.

Tunnelling is not just an idea. There are real-life examples of tunnelling, the earliest known going back to the late twenties when Gamow explained alpha-particle radioactivity. Many heavy nuclei (e.g., radium and uranium) are unstable and seek to go to a (lower) stabler state by shedding an alpha particle — see Fig.7.3. The transformation involved may be symbolically expressed as:

$$_{Z}X^{A} \quad \xrightarrow{\alpha\text{-emission}} \quad _{(Z-2)}Y_{*}^{(A-4)}.$$

Parent nucleus   Daughter nucleus

80 What is reality?

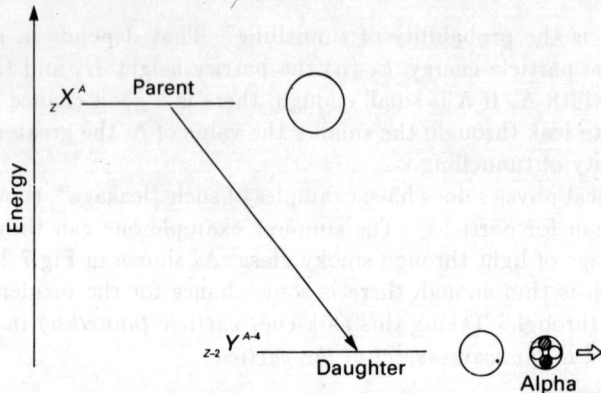

Fig.7.3 Energetics of alpha particle emission. Observe that the energy of the daughter nucleus is lower than that of the parent nucleus.

The emission of the alpha particle from the parent nucleus occurs due to tunnelling — see Fig.7.4. Suppose we now have a large number

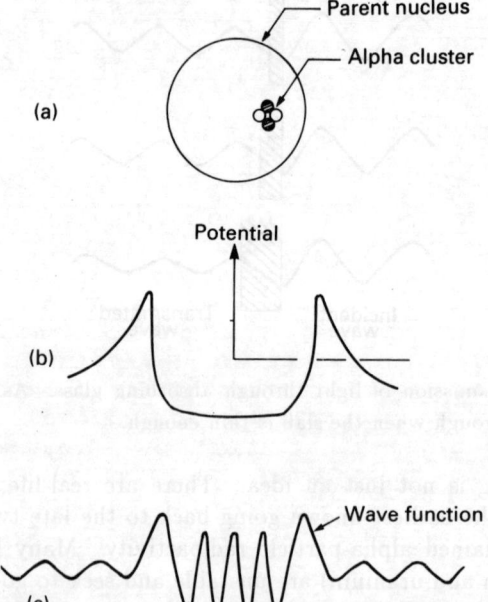

Fig.7.4(a) shows a schematic of the parent nucleus. In it we visualise an alpha particle that is knocking around and trying to escape. (b) shows the potential well-experienced by the alpha particle. It is like a box with walls of finite height and finite thickness. The wave function is shown in (c). As can be seen, there is a strong wave amplitude outside the nucleus. It is this which leads to tunnelling.

N of such radioactive nuclei. As time goes on, these will transform to the daughter species through the emission of alpha particles. If the number of parent nuclei that survive is plotted as a function of time, one would get a curve as in Fig.7.5. The quantity λ identified in that figure is called the radioactive *decay constant*. The smaller the barrier, the larger is the value of λ; correspondingly, the decay is more rapid. In the counter example to ME that has been proposed, one has to deal with the tunnelling of photons which is why I have digressed into a discussion of tunnelling.

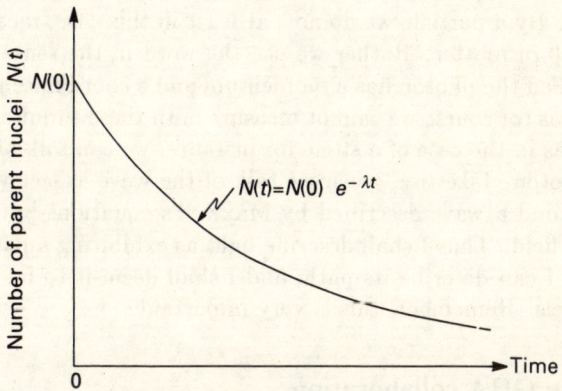

Fig.7.5 The number $N(t)$ of parent nuclei surviving at time t. Observe that the decay is exponential. The constant λ is called the decay constant, and depends on the shape of the potential barrier (Fig. 7.4(b)).

7.3 A word of caution

I must digress once more, but this time about the use of words. I have already touched upon this topic earlier but I need to do so again. The founding fathers of QM have cautioned about the use of words for describing quantum phenomena. We are all macroscopic classical beings, and our normal experience is in the classical domain. In fact, that is the reason why classical physics took shape first — people wanted to understand what they saw and experienced, like why stones fell when dropped, why the sky is blue, etc. Indeed the words and the languages that we use have also been shaped by our experiences. This being the case, it is no wonder that such words fail when we try to describe certain quantum phenomena.

The inadequacy of language is nothing new. Take for example the word *satyagraha*. When Mahatma Gandhi launched the *satyagraha* movement in the early thirties, the English described it as a civil-disobedience movement. This of course is a gross injustice to and distortion of the true meaning of *satyagraha*. So here you have an example of the inadequacy of language even in normal life. There are no problems when we use the language of mathematics for describing physical phenomena but right now I have to use words to describe the tunnelling of photons which is why I am administering some caution!

As you already know, light can behave either as a "particle" or as a "wave". By a particle we do not, at least in this case, mean a hard solid lump of matter. Rather we use the word in the sense that the object called the photon has a momentum and a coordinate just like a particle has (of course we cannot measure both these simultaneously). Further, as in the case of a stone for instance, we can talk of the *path* of the photon. Likewise, when we talk of the wave-aspect we do not have in mind a wave described by Maxwell's equations but rather a quantum field. Thus I shall describe light as exhibiting a particle-like feature if I can describe its path; and I shall deem it to be wave-like if it tunnels. Remember, this is very important.

7.4 The GHA collaboration

In March 1991, Professor Partha Ghose of the S.N. Bose National Centre for Basic Sciences, Calcutta, Dr. Dipankar Home of the Bose Institute, Calcutta and Professor G.S. Agarwal of the Central University, Hyderabad (GHA) published a paper entitled *An experiment to throw more light on light*. Just in case you are confused, I should add that S.N. Bose refers to the discoverer of Bose statistics (see the companion volume *Bose and His Statistics*), the Bose of the Bose Institute refers to Sir J.C. Bose about whom you will find more in Box 7.1. Before I discuss the idea of Ghose, Home and Agarwal, a few words about its genesis.

Box 7.1 Jagadish Chandra Bose was the first Indian to establish an international scientific reputation. He was born in 1858 and received his education in Calcutta. Attracted to a study of physics, he went to England for higher studies. There he earned the D.Sc. degree from the University of London. On return to Calcutta, he became the first Indian to occupy the position of Professor of Physics in the Calcutta University. But because he was not a white man, the Government (remember we were under British rule) paid Bose only a fraction of the salary it paid to Britishers. In protest, Bose refused to draw pay but continued to work. His satyagraha worked!

Bose made pioneering contributions in the area of microwaves. Even prior to Marconi, Bose demonstrated the passage of electromagnetic waves through solid walls. Naturally this made Bose famous. Later he repeated this experiment at the Royal Institution in London before an audience that included Lord Kelvin.

In later years, Bose shifted his interest from physics to plant physiology. He devised ingenious instruments to show that plants respond to various stimuli. In 1917, he founded the Bose Institute. It is located adjacent to the Calcutta University. He was knighted by the British Government in that same year and was elected a Fellow of the Royal Society a few years later.

Bose had strong literary interests and was a good friend of poet Rabindranath Tagore. He died in 1937.

As you know 28 February has been declared as National Science Day (this is in commemoration of the discovery of the Raman effect; for details about this effect, see the forthcoming volume *Raman and His Effect*). The S.N. Bose Centre celebrates this day by holding a day-long seminar on topics in physics. A couple of years ago, one of the speakers was Professor T. Pradhan, formerly of the Saha Institute of Nuclear Physics, Calcutta and the founder of the Institute for Physics, Bhubaneshwar. Professor Pradhan strongly feels that research in India is often imitative, quite unnecessarily. By way of inspiring his audience, he chose to recall an experiment performed a long time ago by J.C. Bose. After the lecture, Home and Ghose began to wonder what would be the outcome if Bose's experiment were to be repeated with a source that emits single photons, i.e., one photon at a time. Soon followed the GHA collaboration.

I should say a few words here about sources of electromagnetic radiation. In the ultimate analysis, electromagnetic radiation must always be described by quantum physics. However, if the source intensity is sufficiently strong, the quantum aspects can be ignored and the electromagnetic field can be adequately described by classical physics, i.e., Maxwell's equations. Roughly speaking, the quantum electrodynamical field is then excited to a very high state. The quantum number involved being very high, classical laws may be expected to apply (thanks to Bohr's correspondence principle—see Part I). In short, where radiation is involved, the thumb rule is: If the number of photons is small, use QED; if the number of photons is large, classical

84 *What is reality?*

electrodynamics can be employed without any danger.

This distinction is emphasised because shortly I shall describe certain (quantum) experiments whose classical analogue was performed a long time ago.

7.5 Experiment of J.C. Bose

We now consider the experiment of J.C. Bose which got Ghose and Home thinking. This experiment was performed way back in 1897. Bose was a pioneer in generating electromagnetic radiation having wavelengths of the order of a few mm. Such radiation is referred to as *millimetre waves*. Experiments in this region are quite difficult but Bose managed many clever ones. A schematic of the one which is of current interest is shown in Fig.7.6.

S is a source of millimetre-wave radiation. In his experiments, Bose employed three different frequencies; a given run or a given series of measurements were performed with radiation of fixed wavelength. On a spectrometer table facing the source was placed a pair of prisms as shown. There was a small gap between the prisms, and the magnitude of this gap could be varied. Beyond the prism arrangement was a detector of electromagnetic radiation. The position of this detector could be varied, and Bose made observations by keeping it either at A or at B (see Fig. 7.6). In the former position one could observe the directly transmitted radiation and in the latter position the reflected radiation.

Bose observed the following:

1. When the two prisms were flush in contact, no radiation was detected at B. Only at A was the radiation detected, implying that a homogenous slab does not reflect but only transmits.
2. When a gap is introduced, some radiation is detected at B provided the gap exceeds a certain critical value. In other words, a small component of the incident radiation is now reflected.
3. As the gap is further increased, the reflected component increases while the transmitted component decreases. Eventually, there is only a reflected component and no transmitted component.
4. As the wavelength of the incident radiation is increased, the size of the gap at which the double-prism arrangement becomes totally reflecting increases.

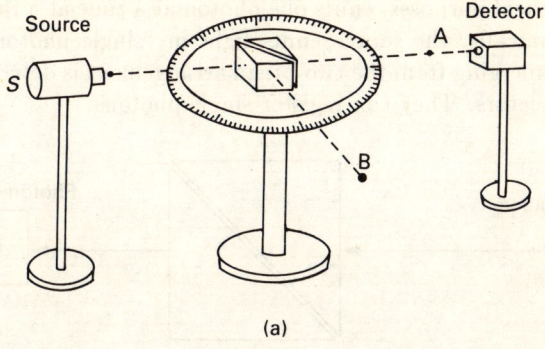

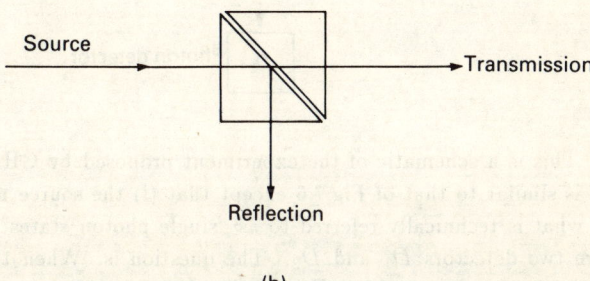

Fig.7.6 Arrangement used by J.C. Bose in his double-prism experiment. (a) shows the overall view while (b) shows the measurement configurations.

In his celebrated *Lectures on Physics* series, Feynman discusses in detail the theory of the experiment of J.C.Bose (vol.II, Chapter 33, p. 12) but does not ever make a mention of Bose; one does not know why. I wonder if Feynman was aware that this experiment was first performed by Bose. I wouldn't be surprised if he did not know about J.C.Bose at all; in fact, few in India know about him today although in his days JCB was quite famous.

7.6 Back to GHA

We now consider the experiment as proposed by GHA, a schematic of which is shown in Fig.7.7. It is similar to the experiment of J.C. Bose except that the source of light is now a photon source, which,

86 *What is reality?*

for all practical purposes, emits one photon at a time at a rather slow rate. Technically, the source emits light in "single photon states". The light emerging from the two-prism arrangement is detected using photon detectors. They can register single photons.

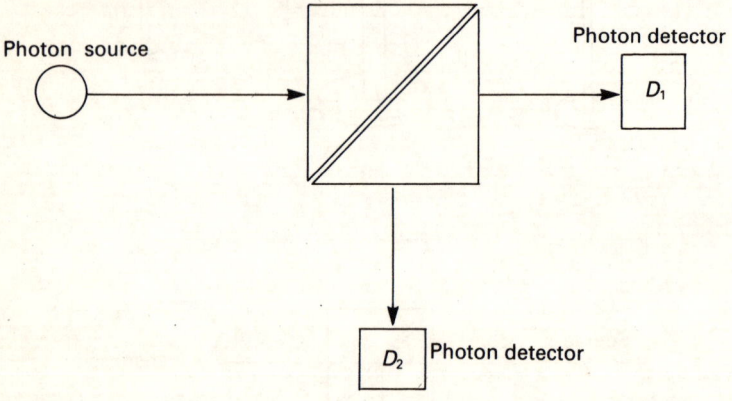

Fig.7.7 This is a schematic of the experiment proposed by GHA. Essentially it is similar to that of Fig.7.6 except that (i) the source now emits light in what is technically referred to as "single photon states", and (ii) there are two detectors D_1 and D_2. The question is: When the source emits a light quantum, do both D_1 and D_2 click or only one of them? If so, which one, D_1 or D_2?

With an arrangement as described, several logical possibilities exist which are listed below:

1. Only one of the two counters D_1 and D_2 clicks at a given time. This means there is detection only at one station. Sometimes it might be counter D_1 and at other times it might be D_2. GHA describe this as detectors "clicking in perfect *anticoincidence*."
2. Both detectors click, i.e., in coincidence.
3. Only D_1 registers, always.
4. Only D_2 registers, always.

I stress again that these alternatives have been written down merely as logical possibilities. The question is: Which one of these is favoured by Nature? GHA argued, on the basis of quantum optics, that possibility 1 is favoured. This has interesting implications.

Firstly, since only one detector registers, we know which way the light went, i.e., whether it went straight (to D_1) or took a turn (and went to D_2). In other words, we know the path taken by light. Since the concept of path is associated with a particle, we take this as an index of particle behaviour. The German word for "which way" is *welcher Weg*; experiments which determine the path are therefore sometimes referred to as welcher-Weg experiments.

Let us go back to GHA. Using QM, they write the state vector of the single photon state emerging from the two-prism arrangement as

$$|\Psi\rangle = \alpha|1,0\rangle + \beta|0,1\rangle \tag{7.1}$$

Here $|1,0\rangle$ and $|0,1\rangle$ refer to states associated with detection by D_1 and D_2 respectively. Observe that (7.1) is a superposed state (like (3.1)). When an experiment is done, *only one of the two possibilities becomes a reality*—the former with a probability $|\alpha|^2$ and the latter with a probability $|\beta|^2 \cdot |\alpha|^2$ gives the transmission probability and $|\beta|^2$ the reflection probability. The important point is that the same experimental arrangement, without any tinkering of any parameter associated with the apparatus, can show the light as *either* transmitted *or* reflected. This is the anticoincidence that GHA refer to. When the photon is transmitted, it tunnels through the gap and light reveals itself as a wave. On the other hand, the fact that detection is by *one* detector only stresses the particle aspect of light because it is indivisible. And the point is that the same arrangement can be used to observe both, the wave and the particle behaviour of light, in contradiction to the mutual exclusivity hypothesis of Bohr.

Notice the following:

1. GHA use QM; nothing more, nothing less. In other words, their prediction follows directly from QM.
2. They exploit tunnelling to highlight the wave aspect instead of interference, as is usually done.
3. QM does not say anything about mutual exclusivity; it was Bohr who brought ME in (as a part of CI).

How do we know that Nature actually behaves the way GHA said it would? GHA declared, "Let experiments decide."

7.7 The actual experiment

The scene now shifts to Japan where Mizobuchi and Ohtaké have been engaged in doing experiments to verify GHA's ideas. Their experimental arrangement is sketched in Fig.7.8. A pulsed laser source together with some gadgetry (which does not concern us) was used for producing light in a single photon state. Of interest to us is their detector set up. The anticoincidence concept is important and

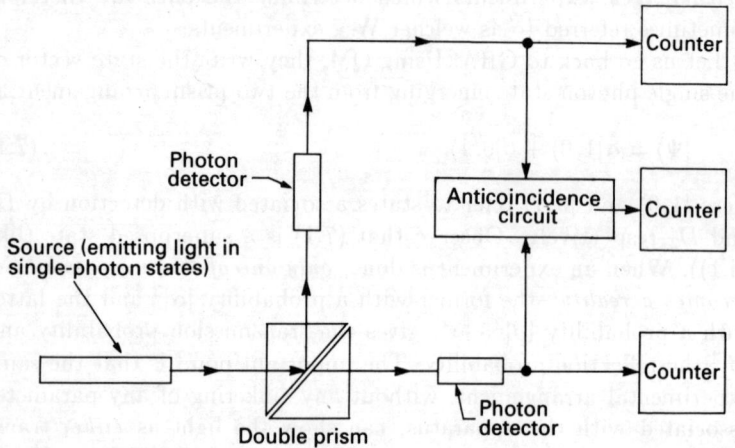

Fig.7.8 Experimental arrangement used by Mizobuchi and Ohtaké to verify the GHA proposal.

is explained in Fig.7.9. If a log of events is made, typically it would appear as below:

Event no.	Detector 1	Detector 2	Anticoincidence
1	✓	×	Yes
2	×	✓	Yes
3	✓	✓	No
4	×	✓	Yes
5	✓	✓	No
6	✓	×	Yes
7	×	✓	Yes

In the above, events 1,2,4,6,7 represent perfect anticoincidence. However, events 3 and 5 represent coincidence. How is this possible? Now perfect anticoincidence is expected only in an ideal experiment. Since no experiment is ideal, stray coincidences may be expected even when the source is emitting single photons.

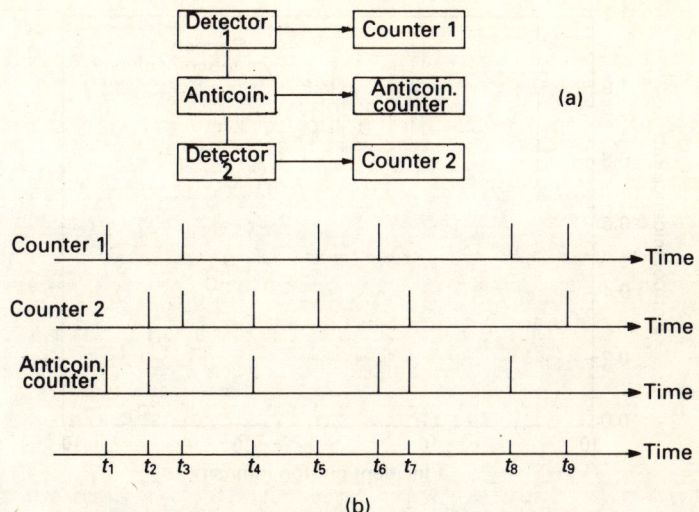

Fig.7.9 The anticoincidence concept. (a) shows the electronics. D_1 and D_2 are the two photon detectors while C_1 and C_2 are counters which count the number of photons detected by the respective detectors. AC is an anticoincidence circuit. This delivers no output pulse when it receives pulses simultaneously at its two inputs. If only one input pulse is received, the anticoincidence circuit delivers a pulse to the counter ACC where it is counted. (b) shows a time chart of the pulses arriving at D_1 and D_2. At times t_3, t_5 and t_9, the AC produces no output. Thus at time t_9, the ACC would have registered only 6 counts.

Turning to the conditions of the experiment, to start with Mizobuchi and Ohtaké arranged so that light has equal probability for transmission as for reflection, i.e., $|\alpha|^2 = |\beta|^2$. Also the rate of emission of photons from the source was kept low. Under these conditions they did see perfect anticoincidence as GHA had predicted, within experimental errors of course.

What happens when the source intensity is substantially increased? The electromagnetic field becomes so strong that it is well described by Maxwell's equations, i.e., classical electrodynamics. And in this situation we already know from the work of J.C. Bose that both detectors would respond.

Figure 7.10 shows the measured anticoincidence probability as a function of the incident light intensity. Here the dotted line represents complete anticoincidence. For low incident intensities, complete anticoincidence is observed exactly as per GHA prediction. And as

90   *What is reality?*

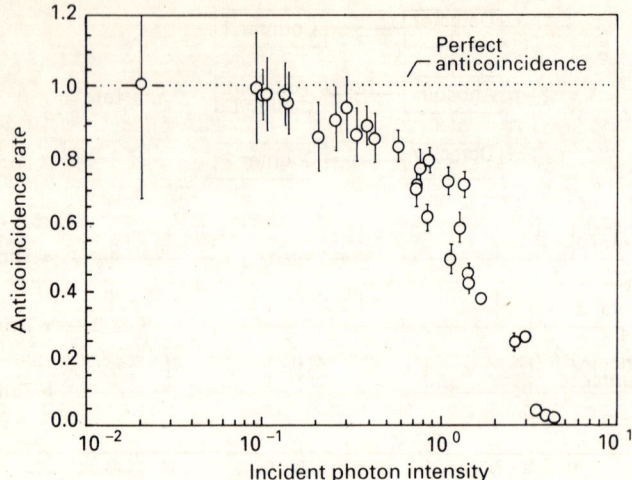

Fig.7.10 Anticoincidence probability as a function of source intensity as observed by Mizobuchi and Ohtaké. At low source intensity, the quantum conditions apply and the probability is unity as predicted by GHA. When source intensity is high, there is a crossover to the classical regime.

the intensity is increased, there is a crossover to the classical regime where the anticoincidence is zero. It should be mentioned that in this series of experiments, the prism gap was kept constant.

Figure 7.11 shows what happens when the source intensity is kept constant (and in the quantum regime) while the gap in the double

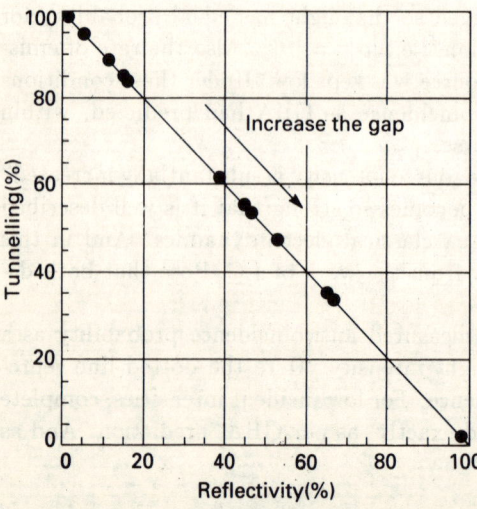

Fig.7.11 The ratio of transmission to reflection, as the gap is varied. This data is due to Mizobuchi and Ohtaké. As was shown by J.C. Bose, transmission dominates when the gap is small and reflection takes over when the gap is large. The significant point is that Bose used a classical source whereas this experiment employed a quantum source.

-prism arrangement is varied. The tunnelling probability then decreases whereas the reflection probability increases. GHA have also pointed out that if the experiment is repeated with a "thermal" source, the results would be in accord with classical theory. This too has been checked by the Japanese workers.

7.8 Concluding remarks

The GHA prediction and its confirmation by the Japanese is only the beginning because many other experiments to check complementarity are now being proposed, using not only photons but also neutrons. More work is therefore expected in this area. As Ghose and Sinha Roy remark:

> The formalism of quantum mechanics allows the possibility of experimental conditions not encompassed by the usual formulation of the complementarity principle with its emphasis on the *mutual exclusiveness* of the conditions which allow unambiguous use of the classical concepts of waves and particles.

I stress again that QM does *not* demand mutual exclusiveness; it is Bohr who does. GHA show that in fact, in the double-prism experiment, QM predicts that both the particle- and the wave-aspects *can* be seen with the same experimental arrangement.

One cannot blame Bohr for not being able to foresee an exception such as the one now discovered. In any case, this sort of failure to foresee all possibilities has happened before (e.g., Newton's inability to foresee that there was a ceiling velocity), and it is likely to happen again. The point simply is that complementarity may not enjoy in future the high pedestal it has been accustomed to for so long. On the other hand, the dethroning might not be before a hot and animated debate. Remember, there is still a strong pro-Bohr lobby!

It is interesting that apart from the idea of complementarity, many of Bohr's other arguments are also beginning to come now under scrutiny. For example, people are saying that the counter argument produced by Bohr to negate Einstein in their famous debate in 1930 is incorrect. Einstein had challenged QM and in its defence Bohr had invoked general relativity. General relativity is not a part of QM, and indeed even today one does not know how to marry it to QM. Under the circumstances, Bohr's appeal to general relativity is "illegitimate". According to these critics, Bohr failed to establish the superiority of quantum mechanics as a valid and self-consistent theory.

I must confess I don't understand all this criticism but I do wonder why these issues were not raised earlier. Why did not Pauli, Dirac and Born, for example, criticise Bohr? Were they all so much pro-QM that they did not even pause to examine whether Bohr was correct in his arguments or wrong? Or was it because of Bohr's father figure? I don't know. Perhaps you can amuse yourself by thinking about it or arguing with your friends!

8 Laboratory Cousins Of Schroedinger's Cats

> One wants to be able to take a realistic view of the world, to talk about the world as if it is really there, **even** when it is not being observed ...
>
> J.S. Bell

The cat paradox discussed in Chapter 3 highlights one of the philosophical issues underlying quantum mechanics, namely, the issue of reality. During the last decade or so, thanks particularly to Leggett, there has been an intensive discussion as to whether a meaningful laboratory experiment can actually be performed which would be the analogue of the cat experiment proposed by Schroedinger. If so, what would be the implications of the possible outcome? Leggett refers to such experimental systems as the "cousins of Schroedinger's cats" while Bell refers to such experiments as "experimental metaphysics".

Before describing these recent developments, I would like to take you back for a moment to the cat paradox. Let $|\Psi_1\rangle$ and $|\Psi_2\rangle$ be the two possible states of the cat, i.e., macroscopic system. According to QM, the final state of the system is the superposed state

$$|\Psi\rangle = a|\Psi_1\rangle + b|\Psi_2\rangle. \qquad (8.1)$$

Note the (macroscopic) system is NOT in a particular macroscopic state but in a *linear superposition* of macroscopically *different states*. As I described in Chapter 3, the problem starts precisely here. From our experience, we can expect the system either to be in the state $|\Psi_1\rangle$ or the state $|\Psi_2\rangle$, but *not* in the superposed state described by (8.1), i.e., we can either have a live cat or a dead cat but not one which is both live and dead.

One way of getting out of this jam is to say: "Don't lose sleep over this problem. No experiment can reveal the difference between (8.1) and a *classical mixture* of $|\Psi_1\rangle$ and $|\Psi_2\rangle$. As far as the latter is concerned, you will see only a live or a dead cat. And in the QM case too you will see the same result, despite (8.1)."

Let me make the above statement a bit more quantitative. Say there is an operator $\hat{\Omega}$. I hope you remember from Part I that an operator is what we use for making a measurement. We denote the outcome of the experiment as $\langle\hat{\Omega}\rangle$ and ask what the outcomes will be when $|\Psi\rangle$ is described by a mixture of $|\Psi_1\rangle$ and $|\Psi_2\rangle$ on the one hand and by a superposition of $|\Psi_1\rangle$ and $|\Psi_2\rangle$ on the other. The answer is:

$$\langle\hat{\Omega}\rangle_{\text{mixture}} = |a|^2\Omega_{11} + |b|^2\Omega_{22}. \tag{8.2a}$$

Here Ω_{ij} is an element of the matrix

$$\begin{pmatrix} \Omega_{11} & \Omega_{12} \\ \Omega_{21} & \Omega_{22} \end{pmatrix} \tag{8.2b}$$

Similarly,

$$\langle\hat{\Omega}\rangle_{\text{superposition}} = |a|^2\Omega_{11} + |b|^2\Omega_{22} + a^*b\Omega_{12} + ab^*\Omega_{21}. \tag{8.2c}$$

Thus what distinguishes $\langle\hat{\Omega}\rangle_{\text{mixture}}$ from $\langle\hat{\Omega}\rangle_{\text{superposition}}$ is the presence of the off-diagonal elements Ω_{12} and Ω_{21} in the latter. It used to be argued that, for technical reasons, Ω_{12} and Ω_{21} vanish in all practical situations. So, as Leggett says, "at first sight, there is simply no quantity which we could measure to distinguish a pure state [i.e., a superposed state] from a mixture".

Leggett then came up with an experiment which appeared to get over this problem. He actually proposed two experiments, one which concerns *macroscopic quantum tunnelling* (MQT) and the other with what is called *macroscopic quantum coherence* (MQC). Consider Fig.8.1(a) which shows the potential well appropriate to a quantum system. Suppose the system is in the state shown by the dot. Over a period of time, the system can leak through the barriers, and there is a rate for this leakage. All this we know already.

Normally, tunnelling occurs only in microscopic systems. You might recall the problem mentioned in Part I of a car tunnelling through a small mound (see Fig.3.11, Part I). In real life, cars don't tunnel through, at least within the lifetime of a human being. Supposing we can have a macroscopic system in which tunnelling *is* possible in a reasonable time. We would then have macroscopic quantum tunnelling.

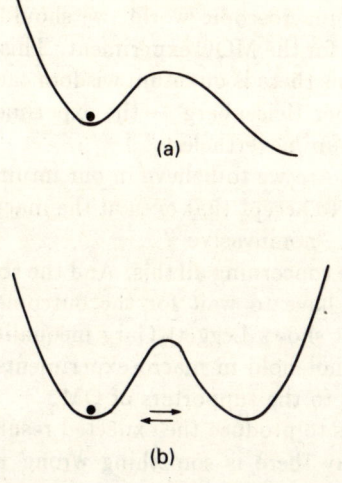

Fig.8.1 (a) and (b) show potentials appropriate to quantum tunnelling leading respectively to leakage and coherent oscillations. The dots represent instantaneous state of the system. In Leggett's experiment, the system is macroscopic. With modern superconducting technology, such a system is conceivable.

The experiment pertaining to MQC is an echo of the ammonia problem discussed in section 3.8 of Part I—see also Fig.8.1(b). The system has two states $|A\rangle$ and $|B\rangle$ and QM predicts the probabilities for these states as (see also Fig.3.9 of Part I)

$$P_A(t) = (1/2)[1 + \cos\omega t] \tag{8.3a}$$

$$P_B(t) = (1/2)[1 - \cos\omega t]. \tag{8.3b}$$

Here there is *coherent* tunnelling, and if such tunnelling is possible in a macroscopic system, we would have MQC. Of the two processes MQT and MQC, let us focus on **MQC as there has been some debate** on it.

In 1985, Leggett and Garg considered MQC and derived an inequality rather like Bell's inequality discussed earlier. In deriving this, they made two assumptions:

1. A macroscopic system with two or more macroscopically distinct states available to it will at all times *be* in one or the other of these states.
2. It is in principle possible to determine the state of the system with arbitrarily small disturbance to it.

We shall refer to these assumptions as A_1 and A_2. These are sometimes referred to as *macro realism* and *non-invasive measurement*.

Both these assumptions are heritages of classical physics and since classical physics works in the macroscopic world, we should expect these assumptions to hold even for the MQC experiment. This is what intuition tells us. As against this there is quantum wisdom cautioning us: "Hey, watch out! Remember Heisenberg — the experiment may be macroscopic but it is quantum nevertheless."

So what are we to do now? Are we to believe in our intuition and swear by A_1 and A_2, or are we to accept that even at the macroscopic level, measurements cannot be "noninvasive"?

There has been much debate concerning all this. And the consensus that has emerged is that we have to wait for the outcome of the experiment. If the experiment shows Leggett-Garg inequality to be violated, then A_1 and A_2 do not hold in macro experiments, which of course would be comforting to the supporters of QM.

What if the experiment fails to produce the expected results? The first tendency would be to say there is something wrong with the experiment. Say the experiment is then very carefully repeated by several groups and they all agree. People would then try to find some way of explaining away the embarassment. Suppose this also is not possible. Leggett says:

> Then we might have to consider very seriously the possibility that the linear laws of quantum mechanics cannot be extrapolated to the macroscopic scale in the way we have always assumed.
>
> Such a conclusion would be of profound significance not only for physics itself, but for the whole of science. For it has been a very deep-seated assumption in the whole of the scientific methodology of the last three hundred years or so that if one can understand the behaviour of the so-called "building-blocks" of matter — be they molecules, atoms, nucleons, quarks or whatever — then one can 'in principle' automatically understand the behaviour of large assemblies of those entities. In other words *complexity* is not taken to be a relevant variable; the behaviour of complex systems is held, without argument, to be determined by the behaviour of the more 'elementary' system composing them. Within such a framework, it would be very difficult, if not impossible, to accommodate the idea that there are corrections to linear quantum mechanics which are functions, in some sense or other, of the *degree of complexity* of the physical system described. Thus, a clear demonstration that corrections of this type exist would pose a severe challenge to some of our most cherished beliefs about the structure of Nature.

One waits with bated breath for the experiments to be performed!

9 Where Does All This Leave Us?

> It [QM] does not really explain things; in fact the founding fathers of quantum mechanics rather prided themselves on giving up the idea of explanation. They were very proud that they dealt only with phenomena: they refused to look behind the phenomena, regarding that as the price one had to pay for coming to terms with nature ...
>
> *John Bell*

9.1 Introduction

I am sure you are rather perplexed by all that you have read so far, and the question that must be uppermost in your mind is the one I have chosen as the title for this chapter. I don't think anyone knows the answer but there certainly are different views on the matter ranging all the way from, "We are back to square one," to "Everything is fine and there is really no problem whatsoever." The best I can do is to give you a cross-section of the prevailing opinion.

Before I get on to describing who is saying what, perhaps we should gather our thoughts.

- In a sense, it all started with EPR.
- They proposed an experiment in which a pair of particles are produced. When the particles are far apart, suitable measurements are made on particle 1. From these measured values, p_2 and q_2 are inferred. Using their definition of reality, EPR then claimed simultaneous reality for p_2 and q_2, in contradiction to QM. This led them to argue that QM was incomplete.
- Bohr dismissed the EPR argument by questioning their definition of reality. He maintained that realities materialise out of potentialities when an observation is made.
- Bohm and Aharonov described a spin version of the EPR experiment, as this was easier to perform.
- Bell considered the spin version of the EPR experiment from a

slightly generalised point of view. He analysed statistical correlations between observations made on the two particles and deduced certain inequalities. Under certain conditions of observation, the inequalities are violated by QM but respected by local realistic theories (i.e., hidden-variables theories).
- A series of experiments consistently showed that Bell's inequalities were violated. According to this, the verdict should go in favour of QM.
- However, objections about "prior conspiracy" between detectors could still be raised in all these experiments.
- So Aspect and coworkers came up with an experiment in which detector conditions were rapidly switched. QM still came out the winner. Local realistic theory could now be totally ruled out.

In 1930, it was more or less Einstein versus all the rest. Perhaps along with Einstein I should also include the names of de Broglie and Schroedinger. But the opposition was small, at least in size. Today the situation is vastly different. True no one any longer says that QM is wrong; on the contrary, everyone accepts QM and in fact uses it in all its glory in day-to-day work. It is still the best prediction machine available. But unlike in earlier years, there are now many shades of opinion concerning the interpretation and indeed also about the need for a change. Figure 9.1 offers a brief glimpse of the spectrum of opinion. Follows now a description of some of these points of view.

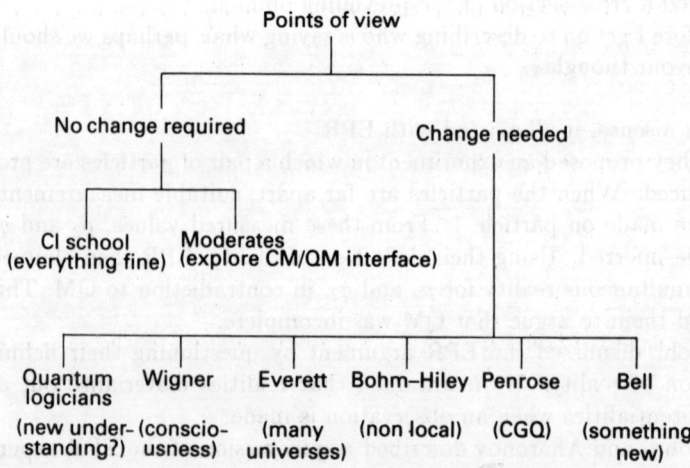

Fig.9.1 Chart showing (partial) spectrum of current attitudes.

9.2 Long live the CI

In politics there is a saying that yesterday's revolutionaries are today's conservatives. Something like that seems to have happened in the case of quantum mechanics also. Its appearance blazed a revolutionary new trail in philosophy. Today the founding fathers are practically all gone, but the school of thought established by them, "the Copenhagen school" if one might call it that, continues to flourish and has a very large following. The very aggressive among the followers say that QM was, is and always will be OK. These supporters don't want any change, and have become arch conservatives. Their view roughly is: "The experiments to test Bell's inequality were not really needed; if anything, they have served to reinforce the supremacy of QM. And there is really no need to break one's head about issues and interpretations; what was good for the founders is good enough for us also." The distinguished physicist Sir Rudolf Peierls goes so far as to object to the use of the term "Copenhagen interpretation" because according to him there is none other.

One of these ardent "loyalists" is Professor John Taylor of King's College, London who believes that we have no choice but to stick to the statistical interpretation. He says:

> We're making a measurement on an aggregate or *ensemble* of identically prepared systems. We thereby obtain a whole set of measurements, one for each of the identical versions of our particular experiment in the ensemble. Hence our results take the form of a probability distribution of particular values for that measurement.

Professor Taylor severely disapproves trying to describe what is going on in an individual system. He cautions:

> We're not allowed to [describe] ... If we take the EPR experiment, which is really the basis of the Aspect experiment, it's clear that a paradox arises there, because we're assuming that when a measurement is made, say, of the spin of a particular particle, we can also measure the spin of a far away particle whose properties are correlated according to the usual quantum mechanical ideas. For example, we might find that the particle nearby has a spin pointing up. From that, we conclude that the other particle far away must (if is correlated) therefore have spin down. This would be paradoxical if you believe that you are indeed measuring individual systems because it would seem that you are actually able to influence that far away particle, and in some ways determine its spin simply by making a measurement on the nearby particle.

The ensemble interpretation says, however, that we are looking at a whole ensemble of such systems. Some 50% of them may have (when we are measuring them) nearby particles with spin up and far away ones with spin down, while the other 50% have the opposite spins. But we can't say in any particular case what that spin of the far away particle is from the measurement nearby, because we don't know about it; we only know about ensembles of such situations.

Taylor further reminds us that out of quantum theory have emerged quantum field theories (recall Part II). True they are not perfect yet but they are absolutely indispensable for explaining many of the observations of high-energy physics. If we rock QM, much of that edifice would also collapse and we would have nothing else to replace it with. This is what you might call a hard-nosed view, and Taylor makes no bones about the fact that he is a hard-nosed physicist (or an arch conservative if you like!).

9.3 The moderates

This is a cautious group. It does not want any clashes with the CI supporters nor does it want to express approval of the radicals. It says (in effect): "QM works wonderfully well and at the moment there is no need to tinker with it. We need not even worry about the interpretation etc., because there are many things we still do not know concerning the structure of QM and its relationship to classical mechanics (CM). After all, QM without CM is not conceivable. So, must we not try to understand the relationship between the two? At present we know how to bridge classical systems with their quantum counterparts only in some very simple cases. We know how to write the Bohr-Sommerfeld quantisation rules and establish linkages via the correspondence principle [these terms are explained in Part I]. We must know how to do all this for more complex systems, especially those which exhibit *chaos*. May be this exercise will shed more light on tricky questions that at present is producing a lot of sound but not light."

9.4 Quantum logic

The struggle to understand the meaning of quantum mechanics has thrown up a new and interesting idea, based on a discovery by the mathematicians von Neumann and Birkhoff. Before I tell you what

Where does all this leave us? 101

they discovered, I must first say something about the logic that we normally use.

We use what is called *Aristotelean logic* because it was Aristotle who first codified it. George Boole in the last century showed how logic can be expressed in algebraic terms. For this purpose we first need some symbols. Table below collects together the symbols we need.

Operation	Symbol	Comment
Statement	A,B...	Each letter stands for a statement, e.g., A could denote: The earth is round.
Not	$\bar{A}$	The symbol $\bar{A}$ negates the statement represented by A. Thus, $\bar{A}$ stands for: The Earth is NOT round.
And	$\wedge$	The symbol $\wedge$ combines two statements. The meaning of (A$\wedge$B) is that the statement (A and B) is true if and only if A and B are both true.
Or	$\vee$	The symbol $\vee$ combines two statements. The meaning of (A $\vee$ B) is that (A or B) is true if and only if at least one of the two statements A,B is true.

We now come to Aristotle who has three laws:

Law of contradiction:
  No proposition is both True and False.

Law of excluded middle:
  Each proposition is either True or False.

Law of identity:
  Each proposition implies itself.

All this may sound a bit formal but nothing to worry about. Starting from the propositions of Aristotle and using the notation of the above table, we can now make new tables as below. These pertain to combinations of statements and are called *Truth Tables*. Truth tables are very important in electronics and in computers, and a glimpse into this area is given in the appendix.

p	q	$p \vee q$	p	q	$p \wedge q$
T	T	T	T	T	T
T	F	T	T	F	F
F	T	T	F	T	F
F	F	F	F	F	F

In the above, T stands for True and F for False. I think the tables should be easy to follow.

You might think I am drifting away from QM; not really! Just be patient for a couple of minutes. Now one of the laws familiar to us in ordinary algebra is the following:

$$A \times (B + C) = (A \times B) + (A \times C). \tag{9.1}$$

This is called the *distributive* law. The corresponding rule in Boolean algebra(i.e., Aristotelean logic) is:

$$A \wedge (B \vee C) = (A \wedge B)(A \wedge C). \tag{9.2}$$

To give meaning to these symbols, I now identify A, B and C with the following statements:

A: In the double-slit experiment, the electron hits the screen at the point X.
B: The electron has gone through slit 1.
C: The electron has gone through slit 2.

With these meanings for A, B and C, the left-hand side of (9.2) reads:

The electron hits the screen at X *and* the electron has gone through at least one of the two slits.

The right-hand side means that at least one of the following two propositions is true:

The electron hits the screen at X and has gone through slit 1
OR
The electron hits the screen at X and has gone through slit 2.

So far, we have been playing with symbols and statements. Let us now compare these statements with the results of actual experiments on double slits as described in Chapter 1 of Part I. The left-hand side of (9.2) (see also the meaning) describes the situation appropriate to the case where there is no spying. And in this experiment we know that we would see interference fringes.

We now consider the right-hand side of (9.2) and also its meaning. This clearly refers to the case where there is spying. And we know that in this case we get a different result. So in QM the two sides of (9.2) are NOT equal. This is a fact of life, whether we like it or not. The message is:

Aristotelean logic is not able to describe quantum experiments.

Where does all this leave us? 103

What do we do now? We appeal to von Neumann and Birkhoff. They observed that though the mathematics of quantum mechanics does not quite obey Boolean algebra, it nevertheless obeys an algebra of its own. And associated with it is another kind of logic. To be sure, it is different from Aristotelean logic but this other logic is able to describe the quantum world.

In some sense, this situation is similar to what was encountered when general relativity was discovered. Till that time everyone believed that space-time was flat but general relativity showed that space-time was curved. In turn this meant giving up Euclid's geometry and settling for something else. Experts would caution that this analogy is not perfect; I agree but it serves my present purpose.

OK, so the logic underlying quantum physics is different and we have a clue to it from quantum algebra. Some research is apparently going on in this area called *quantum logic* but it is not clear, at least to me, how this is going to help in resolving the many questions facing us. No wonder Bell is a bit sceptical about it. He remarks:

> It is my impression that the whole vast subject of 'Quantum Logic' has arisen in this way from the misuse of a word [measurement].

9.5 Progress through consciousness

The founder of this school is Wigner. He proposed these ideas thirty years ago and there are many who support them even today. Wigner believes that the equations of QM would cease to be linear if consciouness is brought in. Once QM becomes non-linear, it would become possible to describe the so-called collapse via mathematics. Of course, one could think of non-linearities in QM without bringing consciousness into the picture but here I am drawing attention to Wigner's point of view.

Matters seem to rest there, and there is no theory in sight which can be regarded as a viable candidate non-linear replacement for QM which also accomodates consciousness. At present such a theory remains a mere hope.

9.6 The participatory universe

John Wheeler has extended some of the traditional concepts concerning measurement etc., to what is called the *participatory universe*. Recall first the standard CI view, paraphrased by Jordan (see Chapter 1), that "we ourselves produce the results of the measurements."

104 What is reality?

Wheeler takes this one step further. He says that it is not merely a question of "observer-created reality"; rather, the observer *participates* in creating the reality. Wheeler further says that, "In this sense the universe does not sit 'out there' " — see also Fig. 9.2. And he adds that "the past has no existence except as it is recorded in the present."

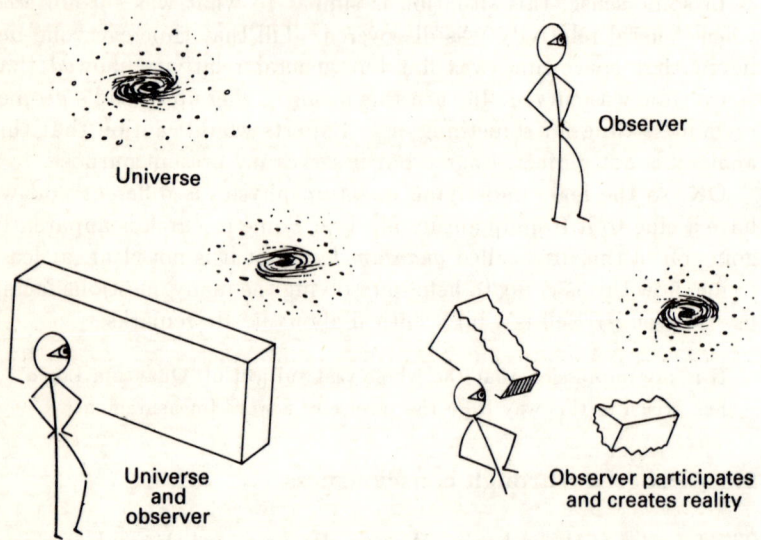

Fig.9.2 According to Wheeler, the observer "participates in the defining of reality." This is illustrated here with a cartoon which is based on one due to Wheeler himself.

I am sure your head must be spinning! So let me make things slightly easier for you by citing the example of the so-called delayed choice experiment, due to Wheeler of course. The idea of this experiment is illustrated in Fig. 9.3. Consider (a). We have here the arrangement needed for the standard double-slit experiment. We now modify the arrangement as in (b). Essentially, at L we have introduced a lens (in place of the film). This lens can combine the light from the two slits and produce two images on the far screen F. For the light source, we use one which emits single photons at a very slow rate. So in this case, we can in principle use ray optics to trace back from the images on F to the slits and identify through which slit the photon came. This would be easier of course with an electronic **array detector** placed at F.

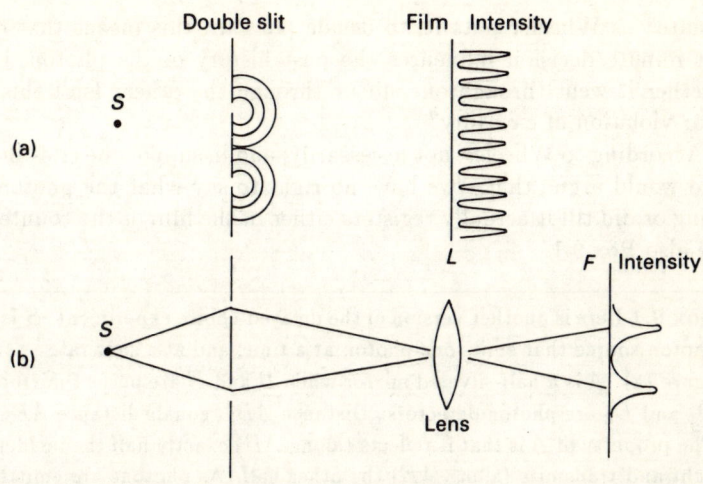

Fig.9.3 Delayed choice in the standard double-slit experiment. (a) shows the standard arrangement. In this one would normally place a film at position L to record the interference fringes. In (b), we have a lens instead of a film. This lens produces images of the slits at F. Whether we have a film at L or at F (or even a photon array detector) is a decision which can be taken after light has passed the double-slit.

The usual double-slit experiment would involve placing a photographic film at L. Even though the source emits single photons, if we wait long enough we would see an interference pattern on the film because every photon "would have passed through both slits".

We now do something very clever. We make the distances slightly large (this is entirely for our convenience) but this in no way disturbs the basic fact that only with two slits and a film we get an interference pattern. Now why this move to increase distances? For the following reason: We wait for the photon to pass through the double slit. *After it has done so, and before it reaches the position L, we decide whether we place a film at L, or we introduce the lens there and place an array detector in position F*. This kind of late decision making is referred to as *delayed choice*.

The late decision does not upset the outcome vis-a-vis QM. As before we get an interference pattern if we do not know which slit the photon went through, which is the case illustrated in Fig. 9.3(a). On the other hand, if we can track the photons as in Fig. 9.3(b), we get a simple superposition of the images of two slits at F. What then is the big deal? Well, simply that "we can wait till the very last

106 What is reality?

minute" as Wheeler puts it, to decide. In turn this means that our last minute decision influences the past history of the photon, i.e., whether it went through one slit or through the other. Isn't this in clear violation of causality?

According to Wheeler, not necessarily; and in support he cites Bohr who would argue that, "we have no right to say what the photon is doing or did till it actually registers either in the film or the counter." See also Box 9.1.

Box 9.1 Here is another version of the delayed-choice experiment. S is a photon source that emits one photon at a time, and at a slow rate — see figure (a). A is a half-silvered mirror while B and E are perfect mirrors. D_1 and D_2 are photon detectors. Distance ABC equals distance AEC. The property of A is that it reflects (along AB) exactly half the incident light and transmits (along AE) the other half. As photons are emitted one by one from S, we find that either the detector D_1 or detector D_2 clicks. Always it is only one detector. We usually describe this by saying that the photon went along $SABCD_1$ (if detection is by D_1) and that it followed the route $SAECD_2$ (if detection is by D_2). The wave function scenario is described in figure (b). Shown here schematically is the wave function at various stages. At stage 1 there is only one peak which is understandable. At stage 2, there are two peaks since light can follow two paths and as yet we do not know which path is being taken. These two peaks are present in stages 3 and 4 also for obvious reasons. At stage 5, the peaks overlap which is understandable.

At stage 6 again we have two peaks which makes sense. But at stage 7

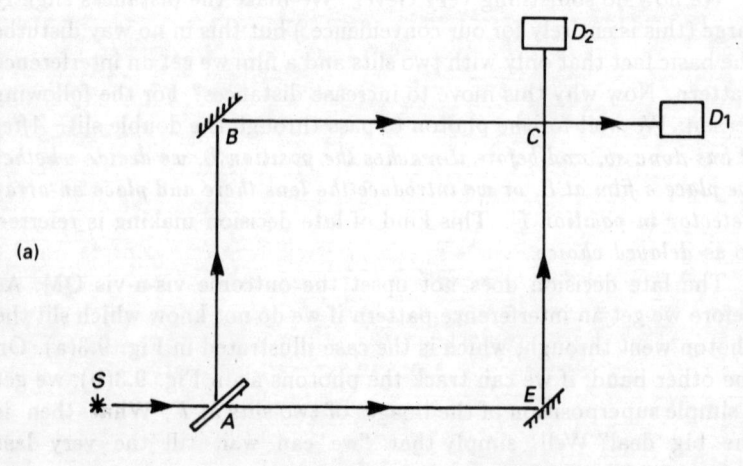

(a)

Where does all this leave us? 107

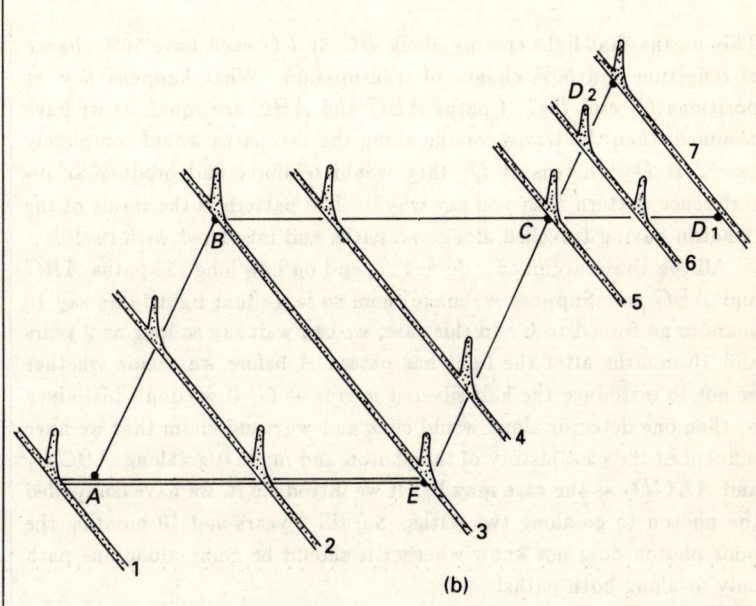

(b)

there is only one peak. In figure (b), the peak is shown a D_2 – this would be the case if D_2 clicks. The double peak of stage 6 collapses to a single peak and using conventional language we would say the photon followed the path $SAECD_2$.

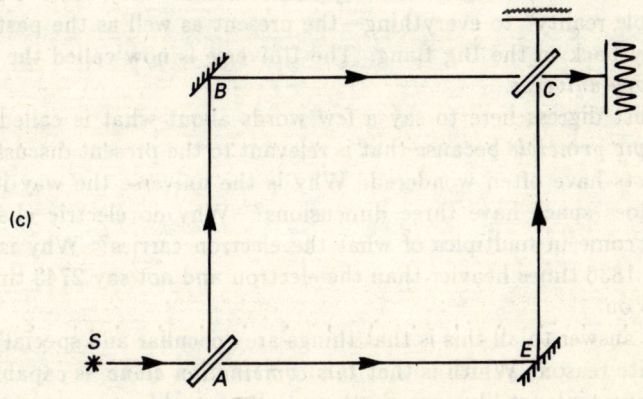

Figure (c) shows a slight variation in the arrangement as compared to (a). At position C, there is now a second half-silvered mirror.

> This means that light coming along BC or EC each have 50% chance of reflection and 50% chance of transmission. What happens now at positions D_1 and D_2? If paths AEC and ABC are equal, as we have assumed, then the waves coming along the two paths would completely cancel at D_2 whereas at D_1 they would reinforce and produce an interference pattern (can you say why?). The pattern is the result of the "photon having travelled along two paths and interfered with itself."
>
> All the above arguments do not depend on how long the paths ABC and AEC are. Suppose we make them so large that light takes say 10 years to go from A to C. In this case, we can wait say as long as 9 years and 10 months after the light has passed A before we decide whether or not to introduce the half-silvered mirror at C. If we don't introduce it, then one detector alone would click and we could claim that we have influenced the past history of the photon and made it go along $ABCD_1$ and $AECD_2$ as the case may be. If we introduce it, we have compelled the photon to go along two paths. So, till 9 years and 10 months, the poor photon does not know whether it should be going along one path only or along both paths!

Why all this peculiar complication? Because from a philosophical point of view, the whole universe can be thought of as the result of a delayed-choice experiment—see Fig. 9.4. Starting from the Big Bang, the Universe cools and expands. Galaxies and stars form, including our own Sun. The latter acquires planets and in the third planet, life evolves. Eventually humans appear and what they observe gives "tangible reality" to everything—the present as well as the past, all the way back to the Big Bang. The Universe is now called the *participatory universe*.

I must digress here to say a few words about what is called the *anthropic principle* because that is relevant to the present discussion. Scientists have often wondered: Why is the universe the way it is? Why does space have three dimensions? Why do electric charges always come in multiples of what the electron carries? Why is the proton 1836 times heavier than the electron and not say 2743 times? And so on.

One answer to all this is that things are "peculiar and special" for a definite reason. Which is that *this combination alone* is capable of producing a planet like our Earth and allowing life to evolve on it.

Where does all this leave us? 109

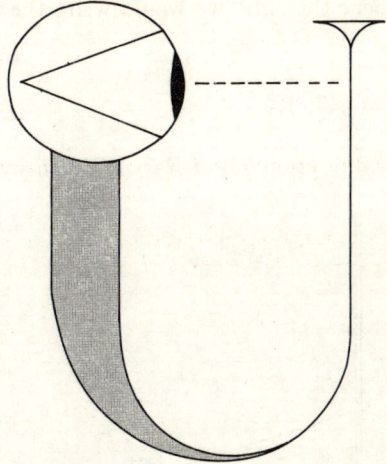

Fig.9.4 According to Wheeler, observer participation gives tangible reality to even the earliest days of the Universe.

Had Nature opted for different physical laws and different values for the fundamental constants, the Universe would have been very different and quite possibly we would not have been around to see it. All this may sound rather egoistic (!) but the argument has been given a scientific dressing. In fact Carter has enunciated a principle called the *anthropic principle* which states:

> What we expect to observe must be restricted by the conditions necessary for our presence as observers.

In a sense, this is similar to what most religions say: "God made the Universe and in particular the world for mankind to inhabit." With all this background, you can see that Wheeler's ideas provide a quantum mechanical backdrop, so to speak, for the anthropic principle.

9.7 The many-universes hypothesis

If you think the participatory universe is weird, wait till you read what I am now about to describe! In 1950, Hugh Everett a Ph.D. student in Princeton University came up with a new interpretation of quantum mechanics in which the wave function does not collapse, i.e., the **R**- process does not occur. Instead, a strange thing happens. Let me try to explain this by considering the double-slit experiment.

What is reality?

If we did not spy near the slits, we would write the wave amplitude as

$$\langle x|1\rangle\langle 1|s\rangle + \langle x|2\rangle\langle 2|s\rangle;$$

see Fig.9.5; also recall section 2.4 of Part I. When we spy, the wave

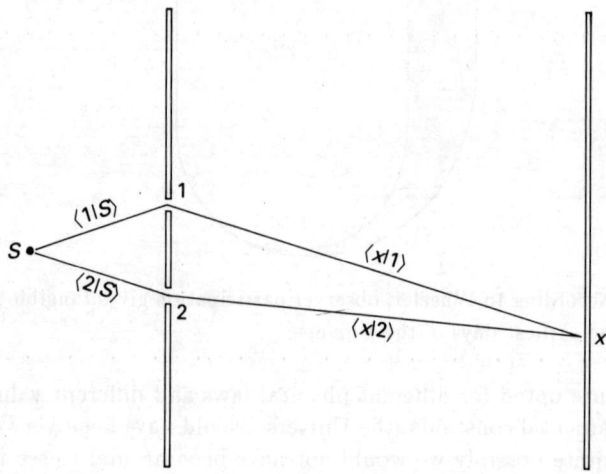

Fig. 9.5 Various partial amplitudes associated with the double-slit experiment.

amplitude changes either to $\langle x|1\rangle\langle 1|s\rangle$ or to $\langle x|2\rangle\langle 2|s\rangle$, depending upon near which slit we spy; this is what we normally refer to as the collapse and this is what gives everyone the creeps. Everett says that nothing like this happens. It is not as if the wave function suddenly becomes truncated and moth-eaten. Instead it just "splits" into two parts one of which remains in our universe and the other wanders off to another universe to oblige another observer spying at the other slit. If you take the cat problem, the wave function is $\{|\text{ alive }\rangle + |\text{ dead }\rangle\}$ till we open the lid. In Everett's picture, the moment we open the lid, the wave function does not collapse but splits into two; if we find that the cat is alive, the part |alive> stays in our universe while the part |dead> appears in another universe. What does that mean? It means that to an observer in that universe, the cat would be dead.

I am sure you would agree that this is crazier than anything that *Alice in Wonderland* has to offer! I wouldn't blame you for thinking so for I am giving a highly diluted (and possibly somewhat distorted) version of Everett's theory. But the essence of it is like what I have described and even experts describe it this way when addressing popular audiences.

Is one supposed to take all this seriously? The answer is "Yes" To reassure you, I must mention that when Everett started trying for an alternative to the well-established CI, he received support from no less a person than Professor Wheeler. Everett published his work in 1957 in the prestigious *Reviews of Modern Physics* and alongside it there was another paper by Wheeler himself, drawing attention to this important work. Inspite of this recommendation, Everett's work remained unnoticed for over ten years until Bryce DeWitt picked up the thread.

Everett's work is not idle speculation; rather it is a serious piece of work based on impeccable mathematics. One thing about mathematical logic is that it is difficult to refute, however unacceptable the result might be psychologically. So people have raised all the inconvenient questions you and I would ask. For example: "People are observing all the time. Each observation represents a measurement. Since this business has been going on for tens of thousands of years (at least), the universe must have split billions of times —see Fig.9.6!

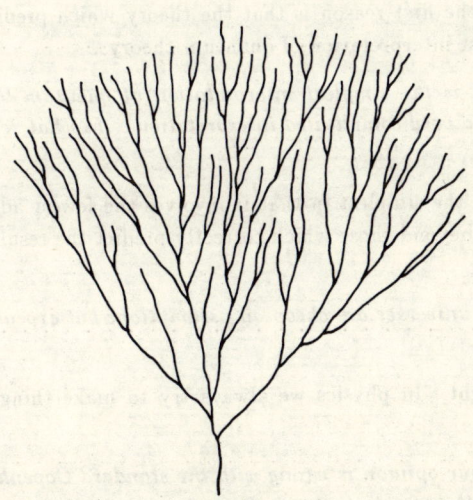

Fig.9.6 Repeated branching of the Universe following observations.

What does the theory have to say about this?" The answer is that the universe does not keep on bifurcating in this fashion. Rather, there are an infinite number of universes even to start with. These exist in a superspace. When the wave function evolves, each universe carries a component. But these components representing alternative "realities" are tied together until we make an observation. The act of observation cuts these mysterious bonds and liberates these alternative realities which then float away in superspace, possibly one component per universe. The observer is left with one of these and he mistakenly thinks that the wave function has collapsed, the rest of the wave function having vanished into thin air. There is no collapse and so there is no problem about how the wave function mysteriously jumps.

I am sure all this would not sound very convincing. Don't worry, lots of people have felt that way. Professor David Deutsch, a theoretical physicist who divides his time between Oxford University and the University of Texas (at Austin), is a strong supporter of the many-universes interpretation. In a radio interview, he was bombarded with many such questions. Here are some excerpts.

Why should we believe in such a monstrous suggestion [about many universes]?

I suppose the first reason is that the theory which predicts them is the simplest interpretation of quantum theory ...

You say it is the simplest interpretation of quantum theory but it seems like a very complicated interpretation ...In what sense is it the simplest?

It is by far the simplest in that it involves the fewest additional assumptions beyond those which correctly predict the results of experiments. ...

So parallel universes are cheap on assumptions but expensive on universes?

Exactly right. In physics we always try to make things cheap on assumptions.

What in your opinion is wrong with the standard Copenhagen interpretation of quantum mechanics?

Well, I've mentioned that the Everett interpretation is more natural in the formal sense. But the best physical reason for adopting the

Everett interpretation lies in quantum cosmology. There one tries to apply quantum mechanics to the universe as a whole ... And immediately one has to stop ... it's logically inconsistent to imagine an observer sitting outside it. Therefore the standard interpretation fails to describe quantum cosmology. Even if we knew how to write down the theory of quantum cosmology, which is quite hard incidentally, we literally wouldn't know what the symbols meant under any interpretation other than the Everett interpretation.

Now in the many-universes interpretation, one seems to hang on to some vestige of objective reality, although it is a multiplied reality.

Yes, that's one of its main advantages.

It is interesting that Wheeler who was one of the early supporters later cooled off as the many-universes interpretation meant carrying "excess metaphysical baggage". However, there are others, like Deutsch, who are not turned off. In fact, Deutsch believes that the many-universes theory can be put to a test!

If we were to plan such an experiment, we would try to think of a situation where standard QM and the many-universes interpretation predict something that is different. But the snag is that we have to deal with observers in these other universes, and we don't know how to do that. However, there is a way out because if we the observers (in our universe) ourselves encountered the different possibilities in the experiment, these possibilities would be registered differently in our memory. So we must dream up an experiment in which there is an *interference effect between these different states of the memory*. Notice that I am essentially talking of *quantum effects in the brain!* Unfortunately, we know so little about the brain that to plan an experiment based on its quantum features is far out in the future. But we don't really need a human brain. After all, don't we use inanimate apparatuses in QM? So why not look for an electronic memory instead? These things are already being used in computers.

The game now becomes very interesting. Unfortunately, "electronic quantum memories" of the type needed for such experiments have not yet been developed but with technology galloping so rapidly, who knows one of these days the kind of experiment I am talking about might in fact be feasible.

Compared to the supporters (most of whom seem to be cosmologists), the objectors seem to be larger in number. Bell, for example, says,

It is extremely bizzare, and for me that would already be enough reason to dislike it. The idea that there are all those other Universes which we can't see is hard to swallow. But there are also technical problems which people usually gloss over ... The actual point at which a branching occurs is supposed to be the point at which a measurement is made. But the actual point at which the measurement is made is totally obscure.

Bell argues, with justification, that high-energy physics experiments often take months and months. At which precise instant during this long period does the universe branch out?

Peierls is another of those who refuses to buy the many-universes idea. He says:

> That's making things unnecessarily complicated. Since we have no means of seeing or even communicating with the other universes, why invent them?

But he is prepared to make a concession. He argues that standard QM gives a list of various possibilities (each with its own probability of occurrence), out of which one is realised when a measurement is made. He then remarks:

> So, in a sense, you can say that quantum mechanics can be represented as a dictionary listing all possible outcomes of all possible initial conditions. Now if instead of "dictionary" you simply use the term "many universes", then we are in line with Everett ... I prefer to use the word "possibilities" or "dictionary of possibilities" rather than "universes".

9.8 The Bohm–Hiley non-local theory

As I mentioned earlier, the Aspect experiment rules out any hidden-variables theory based on objective reality and local causality. What about a non-local theory? Such a theory has been proposed by Bohm who revived an old idea of de Broglie. Way back in 1926 when the founding fathers were agonising, "wave *or* particle?", de Broglie said, "wave *and* particle".

According to de Broglie a particle is *always* accompanied by a wave — Bell calls this a pilot wave. This pilot wave dictates the path of the particle, steering the latter to regions where the wave amplitude is large and keeping it away from regions where the amplitude is small.

Unfortunately, nobody took de Broglie seriously, barring Einstein that is. Naturally, de Broglie was much discouraged. Bell comments:

> De Broglie was laughed out of court in a way that I now regard as disgraceful because his arguments were not refuted, they were simply trampled on.

Bohm resurrected the theory in 1952. Further, he extended de Broglie's ideas to the case of many particles. The theory has been improved even further by Bohm and Hiley. Normal QM has only the wave function ψ. The Bohm–Hiley theory has not only ψ but also the particle coordinate $\mathbf{x}$. Unlike ψ, $\mathbf{x}$ is a classical variable. The time evolution of ψ is described as usual by the (time-dependent) Schroedinger equation. The variable $\mathbf{x}(t)$ obeys the equation

$$\dot{\mathbf{x}}(t) = (1/m)\frac{\partial}{\partial \mathbf{r}} \operatorname{Im} \log \psi(t,\mathbf{r})\big|_{\mathbf{r}=\mathbf{x}} \qquad (9.3)$$

The quantity $\operatorname{Im} \log \psi$ acts like a field and produces an effect on the particle. This field (or if you like potential) is similar in some respects to the other potentials we are familiar with like the electrostatic potential, for example. But there is also an important difference in that this "quantum" potential does not result in a physical force on the particle. The electrostatic potential produces a force and in fact if the strength of the potential increases, so does the force exerted on a charge.

The quantum potential of Bohm merely transmits information. For this reason it is better referred to as an "information" potential. Suppose someone in a crowd shouts, "fire" or "snake"; you know what will happen. The Bohm potential acts somewhat in this manner. If you like, it steers the particle rather in the manner lighthouses guide ships and sophisticated landing systems guide approaching aircrafts in modern airports.

A brief look at the familiar two-slit experiment (recall Part I) would give you some idea of how this whole thing works.

CASE I: No spies

The wave amplitude is described by $\psi(t,\mathbf{r})$ and the particle coordinate by $\mathbf{x}(t)$. The wave influences the particle via (9.3). For every particle coming from the source, there is an accompanying wave. The wave always goes through both the slits but the particle goes through only one. However, the particle is guided by the wave towards places where

116 What is reality?

$|\psi|^2$ is large and away from places where $|\psi|^2$ is small. Remember that one particle does not produce an interference pattern; it is the result of the experiment being repeated many many times. Contact is thus maintained with standard QM. Bell adds:

> It is the de Broglie–Bohm variable **x** that shows up each time. That **x** rather than ψ is legitimately called a "hidden" variable is a piece of historical silliness.

CASE II: Spies present

Detectors D_1 and D_2 are placed near slits 1 and 2 to do the spying. Detector states are denoted by $D_i(t; \mathbf{r}; \text{Yes/No})$. Here $\mathbf{r}_i$ is the position of detector i. The yes/no offers an indication of whether the detector has registered or not.

At $t = 0$, the wave and the particle emerge from the source.

$$\Psi(0) = \psi(t = 0, \mathbf{r}).D_1(0; \mathbf{r}_i; \text{No}).D_2(0; \mathbf{r}_2; \text{No}).$$

For $t > 0$, we write $\Psi(t)$ as

$$\Psi(t) = \Psi_1(t) + \Psi_2(t).$$

When the particle has passed the slits, only one detector would have registered. Accordingly, we have

$$\Psi_1(t) = \psi_1(t.\mathbf{r}).D_1(t; \mathbf{r}_1; \text{Yes}).D_2(t; \mathbf{r}_2; \text{No})$$

$$\Psi_2(t) = \psi_2(t, \mathbf{r}).D_1(t; \mathbf{r}_1; \text{No}).D_2(t; \mathbf{r}_2; \text{Yes}).$$

The two components ψ_1 and ψ_2 are obtained via the equations of the theory. Depending on which detector has registered, one of the Ψ_i's will drop out and the remaining one will guide the particle. When one makes repeated tries, the particle would be observed to go sometimes through slit 1 and at other times through slit 2. In every case, there is a wave contribution from only one slit; naturally the interference pattern is destroyed.

This brings us to the non-local aspect of the Bohm potential. In the case of the electrostatic potential, if the source of the potential is disturbed the effect is felt at a point distance r at time (r/c) later— this of course is consistent with relativity. The Bohm potential on the other hand, acts everywhere instantly. Naturally this arouses suspicion, conditioned as we have been by relativity for nearly ninety years. But Bohm is not so strongly attached to locality and to relativity. He says:

> I would be quite ready to relinquish locality; I think it is an arbitrary assumption. I mean in the last few hundred years it has been given tremendous weight. If you went back 1000 or 2000 years, almost everybody was thinking non-locally.

Wouldn't non-locality produce horrendous paradoxes? Bohm does not think so. On the other hand, he believes that an overdose of relativity has produced a closed mind. This is not to say that relativity must be abandoned. Just that there might emerge a theory for which relativity is a limiting case of some kind. Haven't we gone through this kind of thing before? Why should it not happen one more time and why should we put the stamp of finality on relativity?

Naturally one would wonder about experiments which would reveal deficiencies in relativity or better still, provide a clue as to how non-local effects operate (essentially how particles are "informed" instantly by the wave). Bohm believes that we should not rule out the possibility of such experiments in the future. After all, the concept of the atom was proposed even in ancient times but over two thousand years had to pass before one could perform meaningful experiments at the level of atoms. So maybe we have to be patient and wait.

Obviously, it is not easy to get away with remarks like these, and a number of questions have been fired at the authors. For example: Does not this non-local potential imply in some sense sending signals faster that light? Would the Bohm theory say that something like this is happening in Aspect's experiments? And the answer is, no. Hiley amplifies:

> In Aspect's experiment, although the quantum potential shows there is an instantaneous connection, when we look at the statistical properties of the particles at each end of the connection, they (the particles) appear independent; it is only in the correlations that we see non-locality. It's not clear to me that those correlations can ever be transformed into a signal that makes things go backwards in time [recall the discussion given in *At the Speed of Light* on the violation of causality if signals can travel faster than light].

Bohm is quite aggressive in stating his point of view which obviously is not widely accepted. Nevertheless, he needs to be listened

118 What is reality?

to with some respect. He argues forcefully that QM is nothing but a calculus or a set of rules and does not *explain* anything. Today if we wondered why apples fall and ask what QM has to say about it, QM would give (according to Bohm),

> the statistics of the number of apples arriving in certain places. Now this is similar to the insurance company saying we have statistics on how many people of a certain category will die in a certain year, and that is all we care about! But that is not an explanation.

Bohm also disapproves of the excessive importance given to experiments and letting them decide the course of science (interestingly, the supporters of String theory often use a somewhat similar argument in their debates with the more traditional high-energy physicists). According to him, physics did not begin with experiments. Rather, people began doing experiments when they started asking questions. He laments that philosophy which means love of wisdom does not find a serious place in physics. Philosophy and speculation have been excluded and there is excessive reliance, according to him, on mathematics. It seems to have become the only insight. He adds:

> Nobody minds as long as it's mathematics. People believe that mathematics is truth, but anything else is not.

Finally he pleads:

> But if you will allow mathematical elegance, will you not allow elegance in the conception?

I know all such remarks would be like red rag to hard-core and hard-nosed theoretical physicists but I thought you should not be slanted too much towards one side. Also, you should have a feel for what others think, which is why I have done a bit of "advertisement" for Bohm's views.

9.9 Progress through CQG

Roger Penrose is one of the votaries of change. He has extensively aired his views in his bestseller *The Emperor's New Mind* (claimed to be a popular book but is definitely NOT!). Penrose is unhappy that whereas QM has a grand machinery for dealing with the U-process, it leaves us guessing when it comes to the **R**-process — recall also Fig.3.2. He complains:

> The rules of quantum mechanics appear even to insist that cricket balls and elephants ought to behave in this odd way, where different alternative possibilites can somehow "add up" in complex number combination. However, we never actually *see* cricket balls or elephants superposed in this strange way. Why do we not?

I am glad Penrose is introducing cricket into QM. At the same time, we should be thankful that balls don't appear double, in superposed state etc. Think of all the problems batsmen, fielders and the poor umpire would have! More seriously, Penrose's complaint is that QM does not give a proper recipe for "how the quantum world merges with the classical". So a "new law" is needed, and Penrose is so bold as to speculate that this new law would even enable us to understand how the mind works!

How does Penrose expect this to happen or in other words, what line of enquiry must one pursue? Frankly, I don't understand all that Penrose says but in brief he believes that progress will come by uniting general relativity with quantum mechanics. Actually, people are already on this job and this subject is called *quantum gravity*. There is as yet no accepted theory of quantum gravity but the approach people are taking is to say, "QM is OK; only general relativity must be modified because it is a classical theory, i.e., it must be quantised. The two must then be neatly patched together." Penrose does not quite buy this line and seems to feel that QM too would have to be modified before the merger of these two giants takes place. Penrose refers to this dream theory as *Correct Quantum Gravity* (CQG).

As you know, gravitational effects can be explained in terms of a gravitational field. In Einstein's general relativity, this field is a classical field but in CQG it would be a quantised field like the quantised radiation field. And just as there is a quantum of radiation namely the photon, there would be associated with the quantised gravitational field the quantum of gravitation, i.e., the *graviton*. Penrose identifies what he calls the one-graviton level, and important things are expected at this level. He says that

> as soon as this level is reached, the ordinary rules of linear superposition, according to the **U**-procedure become modified when applied to the gravitons, and some kind of time-asymmetric "non-linear instability" sets in.

Penrose argues that thanks to this non-linearity, the **R**-process takes

120 *What is reality?*

over. In other words, the **R**-process "occurs spontaneously and in an entirely objective way." There it is, a lot of high-powered hand waving!

9.10 Parting thoughts

And so we come to the end of our journey. This time it has been longer than usual—through three volumes as a matter of fact. I have served as your guide and together we have covered a lot of territory. As I take leave, you might wonder what I personally think about the arguments going on. Well, I feel rather like a spectator watching a World Cup Final! Out there in the field, the giants are slugging it out so to speak.

What about the outcome? I can only guess and obviously my guess is as good as yours. I cheer for Bohm for stressing philosophy (I am a bit like that myself) and I agree with Bell who says that it is very probable that the solution to our problem will come through the back door; some person who is not addressing himself to these difficulties will probably see the light.

Finally, I am betting that QM would one day have to yield place to something else, as Penrose and Bell have speculated. Let me therefore conclude with the following words of Bell:

> Quantum mechanics is, at the best, incomplete. We look forward to a new theory which can refer meaningfully to events in a given system without requiring 'observation' by another system.

Einstein would be happy to hear that!

> **Box 9.2** In the early thirties, Einstein and poet Rabindranath Tagore had a meeting. An excerpt from their conversation.
>
> > EINSTEIN: I cannot prove that scientific truth must be conceived as a truth that is valid independent of humanity; but I believe it firmly. I believe, for instance, that the Pythagorean theorem in geometry states something that is approximately true, independent of the existence of man. Any way, if there is a 'reality' independent of man, there is also a truth relative to this reality; and in the same way the negation of the first engenders a negation of the existence of the latter ... Even in our

everyday life, we feel compelled to ascribe a reality independent of man to the objects we use. We do this to connect the experiences of our senses in a reasonable way. For instance, if nobody is in this house, yet that remains where it is.

TAGORE: It is not difficult to imagine a mind to which sequence of things happen not in space but only in time like the sequence of notes in music. For such a mind its conception of reality is akin to the musical reality in which Pythagorean geometry can have no meaning. There is the reality of paper, infinitely different from the reality of literature. For the kind of mind possessed by the moth which eats that paper, literature is absolutely non-existent, yet for Man's mind literature has a greater value of truth than the paper itself. In a similar manner, if there be some truth which has no sensuous or rational relation to human mind, it will ever remain as nothing as long as we remain human beings.

Appendix to Chapter 9

Digital electronics on which the entire computer industry is based (—I am leaving out analog computers which have been more or less eclipsed), revolves entirely around Boolean algebra. In digital electronics, one deals with two states labelled 0 and 1. These states could refer to two different voltage levels, say as in figure (a).

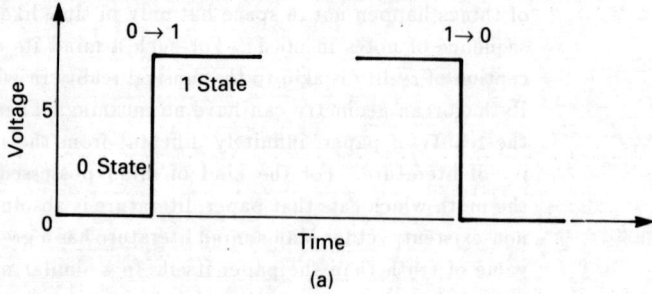

Digital electronics involves circuits which basically deal with elements which can exist in these two states 0 and 1. There are five elementary circuits out of which all complex digital circuits are built. These are shown in figure (b). These are also sometimes called gates.

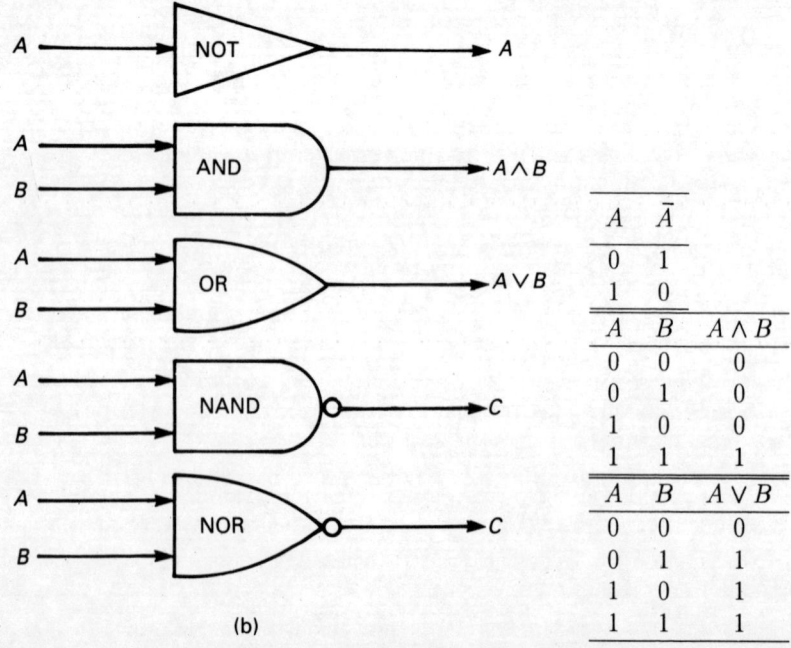

Where does all this leave us?

A	B	$C = \overline{A \wedge B} = \bar{A} \vee \bar{B}$
0	0	1
0	1	1
1	0	1
1	1	0

A	B	$C = \overline{A \vee B} = \bar{A} \wedge \bar{B}$
0	0	1
0	1	0
1	0	0
1	1	0

Alongside each circuit is given the corresponding truth table. There are two other modules or gates which must also be considered. These are shown in figure (c).

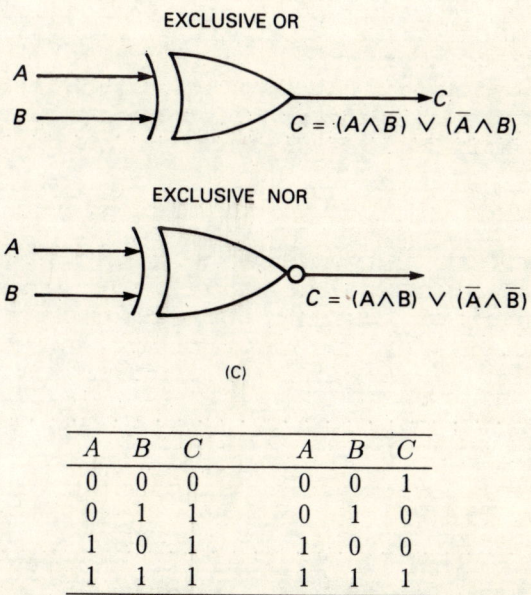

(c)

A	B	C	A	B	C
0	0	0	0	0	1
0	1	1	0	1	0
1	0	1	1	0	0
1	1	1	1	1	1

In every case, the left-hand side of the circuit shows input signals and the right-hand side the output signals. The output state depends on the input states.

124   What is reality?

The heart of a personal computer is a chip called the CPU. It packs close to a hundred thousand gates like those illustrated. While the CPU is working, various elements inside keep switching states from 0 to 1 or from 1 to 0. The gates are used for piping signals, combining them etc. In the fastest microprocessors of today, nearly a million switchings occur every second, if not more.

Suggestions for further reading

The matter dealt with here is in fact quite complicated. It is therefore somewhat difficult to recommend other books which are in the same league as the present one. The best I could come up with are:

1. Herbert, N. *Quantum Reality (Beyond the New Physics)*, Rider: London, 1985.

2. Davies, P.C.W. and Brown, J.R. *The Ghost in the Atom*, Cambridge University Press: Cambridge, 1986.

A few quotes from the above volume have been included in this book.

3. Zohar, D.*The Quantum Self*, Flamingo: London, 1991.

Index

anthropic principle, 108
anticoincidence, 88,(89)
Aristotelean logic, 101,102
Aspect, Alain, 51
Aspect, Dalibard and Roger, (63)
Aspect, Grangier and Roger experiment, 58,(58),(59),65

Bell, J.S., 28,30,45,(46),46,(49),
    67 - 71,75,93,97,103,
    113 - 116,120
Bell's inequality, 47
Bertlmann's socks, 39
Bohm–Aharonov proposal, 32,(41)
Bohm, David, 1,34
Bohm–Hiley non-local theory,
    114 - 116
Bohr, Niels, 12,13,19,21,77
Boolean algebra, 102,103
Bose, Jagdish Chandra, 82-84,(85),(90)

calcium atom, 54,(54)
Carroll, Lewis, 22
Carter, 109
CI (Copenhagen interpretation), 9,99
classical particle (1-D oscillator), (78)
classical physics, 3,4,11
Clauser, 55
complementarity principle, 76, 77,91
complexity, 95
correct quantum gravity (CQG), 118
correlations, 39

de Broglie, Louis, 76,114
decay constant, 81,(81)
delayed-choice, 105,(105)

D'Espagnat, 67
determinism and causality, 3
Deutsch, David, 112,113
digital electronics, 122
distributive law, 101
double-slit experiment, 105,(105),(110)

Einstein, Albert, 11,15,20
Einstein separability, 37,62
epistemology, 2
EPR
    collaboration, 12
    experiment, 14,(32),37,38,(51),97
    objective, 13
    paradox, 12,17
Everett, Hugh, 109,111

Freedman–Clauser experiment, 54,(54),(55)
Fry and Thompson, 57,(57)

gedanken, 11
general relativity, 1,91
Ghosh, Home and Agarwal collaboration (GHA), 82,85,(86),87,(90),91
Gribbin, John, 66

Heisenberg, 72
hidden-variable theory, 31,(44)
Holt and Pipkin, 57

infinite regression, 70

Jordon, Pascual, 9

Note: Page numbers in brackets refer to illustrations

Lamehi-Rachti and Mittig experiment, 63,(64),(65)
law of contradiction, 101
law of excluded middle, 101
law of identity, 101
Leggett, 94
local-realistic theory (LRT), 40

macro realism, 95
macroscopic quantum coherence (MQC), 94,95
macroscopic quantum tunnelling (MQT), 93,(94)
many-universe hypothesis, 109
mercury atom, 55,(56)
Mermin, David, 41
Mizobuchi and Ohtake, 87,(88),(90)
mutual exclusivity (ME), 77

Neumann, 70
non-invasive measurement, 95
nonseparability, 59

ontology, 2

optical commutator, (61)

Pagels, Heinz, 37
participatory universe, 103,108

Penrose, Roger, 28,118,119
photon correlation experiment, 52,(53)
photon experiment, (7)
polarisation, (48)

quantum mechanics, 5,6,11
measurement, 3,6,(8)

Rajaraman, R., 47
Raman–Nath cell, (61),62

Schroedinger, 22,(23),25
cat experiment, 23
wave function, 25
scientific theory, 1

Tagore, Rabindranath, 120
Taylor, John, 99
tunnelling, 77,79,(80)

von Neumann and Birkhoff, 100,103

wave function collapse, 28,(29)

Wheeler, John, 70,75,103,(104),(109),111

Wigner, Eugene, 72,73